ESSAYS ON SCIENCE
by Isaac Asimov

From The Magazine of Fantasy and Science Fiction

FACT AND FANCY
VIEW FROM A HEIGHT
ADDING A DIMENSION
OF TIME AND SPACE AND OTHER THINGS
FROM EARTH TO HEAVEN
SCIENCE, NUMBERS, AND I
THE SOLAR SYSTEM AND BACK
THE STARS IN THEIR COURSES
THE LEFT HAND OF THE ELECTRON
ASIMOV ON ASTRONOMY
THE TRAGEDY OF THE MOON
ASIMOV ON CHEMISTRY

From Other Sources

ONLY A TRILLION
IS ANYONE THERE?
TODAY AND TOMORROW AND . . .

ASIMOV ON CHEMISTRY

ASIMOV ON CHEMISTRY

ISAAC ASIMOV

MACDONALD AND JANE'S · LONDON

First published in Great Britain in 1975
by Macdonald and Jane's
(Macdonald and Company (Publishers) Ltd.)
Paulton House
8 Shepherdess Walk, London N1 7LW

Printed and bound in Great Britain by
REDWOOD BURN LIMITED
Trowbridge & Esher

ISBN 0 356 08297 0

To Carl A. Smith,
easily the best surgeon in the world.

All essays in this volume originally appeared in *The Magazine of Fantasy and Science Fiction*. Individual essays were in the following issues:

CONTENTS

INTRODUCTION

Back in 1959 I began writing a monthly science column for *The Magazine of Fantasy and Science Fiction*. I was given carte blanche as to subject matter, approach, style, and everything else, and I made full use of that. I have used the column to range through every science in an informal and very personal way so that of all the writing I do (and I do a great deal) nothing gives me so much pleasure as these monthly essays.

And as though that were not pleasure enough in itself, why, every time I complete seventeen essays, Doubleday & Company, Inc., puts them into a book and publishes them. As of this moment I have had published ten books of my *F & SF* essays, containing a total of 170 essays. An eleventh is, of course, in the works.

Few books, however, can be expected to sell indefinitely; at least, not well enough to be worth the investment of keeping them forever in print. The estimable gentlemen at Doubleday have therefore (with some reluctance, for they are fond of me and know how my lower lip tends to tremble on these occasions) allowed the first five of my books of essays to go out of print.

Out of *hardback* print, I hasten to say. All five of the books are flourishing in paperback so that they are still available to the public. Nevertheless, there is a cachet about the hardback that I am reluctant to lose. It is the hardbacks that supply the libraries; and for those who really want a permanent addition to their large personal collections of Asimov books* there is nothing like a hardback.

My first impulse, then, was to ask the kind people at Doubleday to put the books back into print and gamble on a kind of second wind. This is done periodically in the case of my science fiction books, with success even when paperback editions are simultaneously available. But I could see that with my essays the case was different. My science fiction is ever fresh, but science essays do tend to get out of date, for the advance of science is inexorable.

And then I got to thinking. . . .

I deliberately range widely over the various sciences both to satisfy my own restless interests and to give each member of my heterogeneous audience a chance to satisfy his own particular taste now and then. The result is that each collection of essays has some on astronomy, some on chemistry, some on physics, some on biology, and so on.

But what about the reader who is interested in science, but is *particularly* interested in chemistry? He has to read through the nonchemical articles in each book and can find only four or five, at most, on his favorite subject.

Why not, then, go through the five out-of-print books, cull the chemistry articles, and put seventeen of them together in a volume we can call *Asimov on Chemistry*. Each individual article is old, but put together like that, the combination is new.

So here is the volume. It has two articles from *Fact and Fancy*, seven articles from *View from a Height*, three articles from *Adding a Dimension*, two articles from *Of Time and Space and Other Things*, and three articles from *From Earth to Heaven*. The articles are arranged, not chronologically, but conceptually. As you turn the pages you will find yourself dealing first with inorganic chemistry, then nuclear chemistry, organic chemistry, biochemistry, and geochemistry. The last two articles are of general scientific interest.

Aside from grouping the articles into a more homogeneous mass in an orderly arrangement, what more have I done? Well, the articles are anywhere from eight to fifteen years old and their age shows here and there. I feel rather pleased that the advance of science has not knocked out a single one of the articles here included, or even seriously dented any, but minor changes must be made, and I have made them.

* *Don't tell me you don't have one!*

In doing this, I have not revised the articles themselves since that would deprive you of the fun of seeing me eat my words now and then, or, anyway, chew them a little. So I have made the changes by adding footnotes here and there where something I said needed modification or where I was forced to make a change to avoid presenting misinformation in the tables.

In addition to that, my good friends at Doubleday decided to prepare the book in a more elaborate format than they have used for my ordinary essay collections and have added illustrations to which I have written captions that give information above and beyond what is in the essays themselves.

Finally, since the subject matter is so much more homogeneous than in my ordinary grab-bag essay collections, I have prepared an index which will, I hope, increase the usefulness of the book as reference.

So, although the individual essays are old, I hope you find the book new and useful just the same. And at least I have explained, in all honesty, exactly what I have done and why. The rest is up to you.

ISAAC ASIMOV
New York, February 1974

ASIMOV ON CHEMISTRY

PART I

INORGANIC CHEMISTRY

1

THE WEIGHTING GAME

Scientific theories have a tendency to fit the intellectual fashions of the time.

For instance, back in the fourth century B.C., two Greek philosophers, Leucippus of Miletus and Democritus of Abdera, worked out an atomic theory. All objects, they said, were made up of atoms. There were as many different kinds of atoms as there were fundamentally different substances in the universe. (The Greek recognized four fundamentally different substances, or "elements": fire, air, water, and earth.)

By combinations of the elements in varying proportions, the many different substances with which we are familiar were formed. Through a process of separation and recombination in new proportions, one substance could be converted to another.

All this was very well, but in what fashion did these atoms differ from one another? How might the atom of one element be distinguished from the atom of another?

Now, mind you, atoms were far too small to see or to detect by any method. The Greek atomists were therefore perfectly at liberty to choose any form of distinction they wished. Perhaps different atoms had different colors, or different reflecting powers or bore little labels written in fine Attic Greek. Or perhaps they varied in hardness, in odor, or in temperature. Any of these might have sufficed as the basis for some coherent structural theory of the universe, given ingenuity enough—and if the Greeks had anything at all, it was ingenuity.

But here was where intellectual fashion came in. The Greek specialty was geometry. It was almost (though not quite) the whole of mathematics to them, and it invaded all other intellectual disciplines as far as possible. Consequently, if the question of atomic distinctions came up, the answer was inevitably geometric.

The atoms (the Greek atomists decided) differed in shape. The atoms of fire might be particularly jagged, which was why fire hurt. The atoms of water were smoothly spherical, perhaps, which was why water flowed so easily. The atoms of earth, on the other hand, could be cubical, which was why earth was so solid and stable. And so on.

This all had the merit of sounding very plausible and rational, but since there was no evidence for atoms at all, one way or the other, let alone atoms of different shapes, it remained just an intellectual exercise; not any more valid, necessarily, than the intellectual exercises of Greek philosophers who were nonatomist in their thinking. The nonatomists were more persuasive in their exercises, and atomism remained a minority view—very much in the minority—for over two thousand years.

Atomism was revived in the first decade of the nineteenth century by the English chemist John Dalton. He, too, believed everything was made up of atoms which combined and recombined in different proportions to make up all the substances we know.

In Dalton's time, the notion of elements had changed to the modern one, so that he could speak of atoms of carbon, hydrogen, and oxygen, rather than of atoms of fire and water. Furthermore, a vast array of chemical observations had been recorded during the seventeenth and eighteenth centuries, which could all be neatly explained by an atomic theory. This made the existence of atoms (still unseen and unseeable) a much more useful hypothesis than it had been in Greek times.

But now Dalton was faced with the same problem the Greeks had faced. How was an unseeable atom of one type to be distinguished from an unseeable atom of another type?

Well, the science of 1800 was no longer geometric. It was merely metric. That is, it was based on the measurement of three fundamental properties: mass (commonly miscalled "weight"), distance, and time. These three together were sufficient to deal with the mechanical Newtonian universe.

Consequently, Dalton followed the intellectual fashion of the times and ignored form and shape. All atoms to him were featureless little spheres without internal structure. Instead, he automatically thought of mass, distance, and time, and of these three the most clearly applicable was mass. As part of his theory, he therefore decided that atoms were distinguishable by mass only. All atoms of a particular element had identical mass, while the atoms of any one element had a mass that was different from those of any other element.

Dalton then went on—and this was perhaps his greatest single contribution—to try to determine what those different masses were.

Now there was no question of actually determining the mass of an atom in grams. That was not possible for quite some time to come. Relative masses, however, were another thing entirely.

For instance, hydrogen and oxygen atoms combine to form a molecule of water. (A "molecule" is the name applied to any reasonably stable combination of atoms.) It can be determined by analysis that in forming water, each gram of hydrogen combines with 8 grams of oxygen. In similar fashion, it could be determined that 1 gram of hydrogen always combines with just 3 grams of carbon to form methane. And, sure enough, 3 grams of carbon always combine with just 8 grams of oxygen to form carbon monoxide.

In this way, we are determining "equivalent weights"; that is, the weights of different elements that are equivalent to each other in the formation of compounds. (A "compound" is a substance whose molecules are made up of more than one kind of atom.) If we set the equivalent weight of hydrogen arbitrarily equal to 1, the equivalent weight of carbon is 3 and the equivalent weight of oxygen is 8.

How can this be related to atoms? Well, Dalton made the simplest assumption (which is what one should always do) and decided that one atom of hydrogen combined with one atom of oxygen to form water. If that were so, then the atoms of oxygen must be eight times as heavy as the atoms of hydrogen, which is the best way of explaining why 1 gram of hydrogen combines with 8 grams of oxygen. The same number of atoms are on each side, you see, but the oxygen atoms are each eight times as heavy as the hydrogen atoms.

JOHN DALTON

Dalton, the son of a weaver, came of a Quaker family and was a practicing Quaker all his life. As Quakers, his parents did not register the boy's birth, so the exact date is uncertain, but he is supposed to have been born in Eaglesfield, Cumberland, about September 6, 1766.

The boy left school at the age of eleven and a year later, in 1778, began teaching at a Quaker school and there grew interested in science. His first love was meteorology and, beginning in 1787, he studied the weather with instruments he built himself. He kept

Thus, if we arbitrarily set the atomic weight of hydrogen equal to 1, then the atomic weight of oxygen is 8. By the same reasoning, the atomic weight of carbon is 3, if a molecule of methane consists of one atom of carbon combined with one atom of hydrogen.

The inevitable next question, though, is just this: How valid is Dalton's assumption? Do atoms necessarily combine one-to-one? The answer is: No, not necessarily.

Whereas 3 grams of carbon combine with 8 grams of oxygen to form carbon monoxide, 3 grams of carbon will also combine with 16 grams of oxygen to form carbon dioxide.

Well, then, if we assume that the carbon monoxide molecule is made up of one atom of carbon and one atom of oxygen, and let C represent carbon and O represent oxygen, we can write the molecule

careful daily records of the weather for forty-six years altogether, to the day he died.

He was the first to describe color blindness, in a publication in 1794. He was color-blind himself and the condition is sometimes called "Daltonism." Color blindness is no asset to a chemist and Dalton was a rather clumsy experimenter. (He was also a poor speaker and could not make money lecturing at a time when scientific lectures were all the rage.)

His meteorological observations led Dalton to consider the composition of the air and from that it was but a step to thinking about the properties of gases. It was easy to suppose, from their properties, that gases were composed of tiny particles.

Others had thought so, but Dalton went on to consider that all matter and not gases alone must consist of these small particles. He adopted the old Greek term "atom" for these particles. He decided that the one key fashion in which atoms differed among themselves was their relative weight. He tried to determine these relative weights from chemical data and was the first to try to prepare a table of what we now call "atomic weights." His "atomic theory" was first presented in 1803.

His Quaker beliefs kept him from accepting the natural honors that would have come to him, but in 1832, at Oxford, he did allow himself to be presented to King William IV. He chose to wear his scarlet academic robe, anathema to Quakers—but since he was color-blind it looked gray to him and he wore it. He died in Manchester on July 27, 1844.

of carbon monoxide as CO. But if carbon combines with twice the quantity of oxygen to form a molecule of a substance with different properties, we can assume, on the basis of atomic theory, that each atom of carbon combines with *two* atoms of oxygen to form carbon dioxide. The formula of carbon dioxide is therefore CO_2.

On the other hand, if you reasoned that the molecule of carbon dioxide was CO, then the molecule of carbon monoxide would have to be C_2O. The first alternative, presented in the previous paragraph, happens to be the correct one, but in either alternative we come up against a molecule in which one atom of one element combines with two atoms of another element.

Once you admit that a molecule may contain more than one of a particular kind of atom, you must re-examine the structure of the water molecule. Must it be formed of an atom of hydrogen and one of oxygen, with a formula of HO? What if its formula were HO_2 or HO_4 or H_4O, or, for that matter, $H_{17}O_{47}$?

Fortunately, there was a way of deciding the matter. In 1800, two English chemists, William Nicholson and Anthony Carlisle, had shown that if an electric current were passed through water, hydrogen and oxygen gases were produced. It was quickly found that hydrogen was produced in just twice the volume that oxygen was. Thus, although the ratio of hydrogen to oxygen in water was 1 to 8 in terms of mass, it was 2 to 1 in terms of volume.

Was there any significance to this? Perhaps not. The atoms in hydrogen gas might be spaced twice as far apart as the atoms in oxygen gas, so that the volume difference might have no relation to the number of atoms produced.

However, in 1811, an Italian chemist, Amedeo Avogadro, suggested that in order to explain the known behavior of gases in forming chemical combinations, it was necessary to assume that equal volumes of different gases contained equal numbers of particles. (The particles could be either atoms or molecules.)

Therefore, if the volume of hydrogen produced by the electrolysis of water was twice the volume of oxygen, then twice as many particles of hydrogen were produced as of oxygen. If these particles are assumed to be atoms, or molecules containing the same number of atoms in both hydrogen or oxygen (the latter turned out to be true), then the water molecule contained twice as many hydrogen atoms as oxygen atoms.

The formula for water could not be HO, therefore, but had to be, at the very simplest, H_2O. If 8 grams of oxygen combined with 1 gram of hydrogen, it meant that the single oxygen atom is eight times as heavy as the two hydrogen atoms taken together. If you still set the atomic weight of hydrogen at 1, then the atomic weight of oxygen is equal to 16.

In the same way, it was found eventually that the formula of methane was CH_4, so that the one carbon atom had to be three times as heavy as the four hydrogen atoms taken together. (The equivalent weight of carbon is 3, remember.) Thus, if the atomic weight of hydrogen is 1, then the atomic weight of carbon is 12.

"Avogadro's hypothesis," as it came to be called, made it possible to come to another decision. One liter of hydrogen combined with one liter of chlorine to form hydrogen chloride. It was therefore a fair working assumption to suppose that the hydrogen chloride molecule was made up of one atom of hydrogen and one atom of chlorine. The formula of hydrogen chloride (allowing "Cl" to symbolize "chlorine") could then be written HCl.

The liter of hydrogen and the liter of chlorine contain equal numbers of particles, Avogadro's hypothesis tells us. If we assume that the particles consist of individual atoms, then the number of hydrogen chloride molecules formed must be only half as many as the total number of hydrogen atoms and chlorine atoms with which we start. (Just as the number of married couples is only half as many as the total number of men and women, assuming everyone is married.)

It should follow that the hydrogen chloride gas that is formed has only half the total volume of the hydrogen and chlorine with which we start. One liter of hydrogen plus one liter of chlorine (two liters in all) should produce but one liter of hydrogen chloride.

However, this is not what happens. A liter of hydrogen and a liter of chlorine combine to form *two* liters of hydrogen chloride. The total volume of gas does not change and therefore the total number of particles cannot change. The simplest way out of the dilemma is to assume that hydrogen gas and chlorine gas are not collections of single atoms after all, but collections of molecules, each of which is made up of two atoms.

One hydrogen molecule (H_2) would combine with one chlorine molecule (Cl_2) to form two molecules of hydrogen chloride (HCl, HCl). The total number of particles would not change and neither

would the total volume. By similar methods it could be shown that oxygen gas is also made up of molecules containing two atoms apiece (O_2).

Using this sort of reasoning, plus other generalizations I am not mentioning, it was possible to work out atomic weights and molecular structures for a whole series of substances. The one who was busiest at it was a Swedish chemist named Jöns Jakob Berzelius who, by 1828,

AMEDEO AVOGADRO

Avogadro was born in Turin, then the capital of the small, independent nation of Piedmont, on June 9, 1776. In 1796 he received a doctorate in law and practiced for three years before turning to science.

What interested him was the discovery made in 1802 by J. L. Gay-Lussac that all gases expand to the same extent with rise in temperature. Avogadro suggested that this was so because all gases (at a given temperature) contained the same number of particles per unit volume. This suggestion, called "Avogadro's hypothesis" was advanced in a paper published in 1811. Avogadro was careful to specify that the particles need not be individual atoms, but might be combinations of atoms—which we now call "molecules," a word Avogadro coined.

Avogadro's suggestion was little regarded in the following decades, which was too bad, because without it there was great and continuing confusion between atoms and molecules and because chemists continued to argue endlessly over the molecular weights and formulas of even simple compounds.

It was not until Avogadro's countryman Stanislao Cannizzaro took up the cudgels on his behalf at a great chemical convention in 1860 that the hypothesis came to be accepted. For Avogadro himself that was too late since he had died in Turin on July 9, 1856.

Now, of course, Avogadro is famous. His name is applied to the number of atoms or molecules present in an amount of substance that has a mass of its atomic (or molecular) weight in grams. This number is 602,252,000,000,000,000,000,000 and it is called "Avogadro's number."

had put out a series of atomic weights that were pretty darned good even by modern standards.

However, the course of true love never does run smooth; nor, it seems, does the course of science. A chemist is as easily confused as the next guy; and all during the first half of the nineteenth century, the words "atom" and "molecule" were used interchangeably. Few chemists got them straight, and few distinguished the atomic weight of chlorine, which was 35.5, from the molecular weight of chlorine, which was 71 (since a molecule of chlorine contains two atoms). Then again, the chemists confused atomic weight and equivalent weight, and had difficulty seeing that though the equivalent weights of carbon and oxygen were 3 and 8 respectively, the atomic weights were 12 and 16 respectively. (And, to make matters worse, the molecular weight of oxygen was 32.)

This reduced all chemical calculations, upon which decisions as to molecular structure were based, to sheer chaos. Things weren't too bad with the simple molecules of inorganic chemistry, but in organic chemistry, where molecules contained dozens of atoms, the confusion was ruinous. Nineteen different formulas were suggested for acetic acid, which, with a molecule containing merely eight atoms, was one of the simplest of the organic compounds.

Then, in 1860, a German chemist named Friedrich August Kekule organized the first International Chemical Congress in order to deal with the matter. It assembled at Karlsruhe in Germany.

The hit of the Congress was an Italian chemist named Stanislao Cannizzaro. In formal speeches and in informal talks he hammered away at the importance of straightening out the matter of atomic weights. He pointed out how necessary it was to distinguish between atoms and molecules and between equivalent weights and atomic weights. Most of all, he explained over and over again the significance of the hypothesis of his countryman, Avogadro, a hypothesis most chemists had been ignoring for half a century.

He made his case; and over the course of the next decade, chemistry began to straighten up and fly right.

The result was pure gold. Once Cannizzaro had sold the notion of atomic weights, a few chemists began to arrange the elements in the order of increasing atomic weight to see what would happen. About sixty elements were known in 1860, you see, and they were a bewildering variety of types, makes, and models. No one could predict how many more elements remained to be found nor what their properties might be.

The first attempts to make an atomic weight arrangement seemed to be interesting, but chemists as a whole remained unconvinced that it was anything more than a form of chemical numerology. Then along came a Russian chemist named Dmitri Ivanovich Mendeleev who, in

1869, made the most elaborate arrangement yet. In order to make his table come out well, he left gaps which, he insisted, signified the presence of yet undiscovered elements. He predicted the properties of three elements in particular. Within a dozen years those three elements were discovered, and their properties jibed in every particular with those predicted by Mendeleev.

The sensation was indescribable. Atomic weights were the smash of the season and a number of chemists began to devote their careers to the more-and-more accurate determination of atomic weights. A Belgian chemist, Jean Servais Stas, had, in the 1860s, already produced a table of atomic weights far better than that of Berzelius, but the matter reached its chemical peak in the first decade of the twentieth century, just a hundred years after Dalton's first attempts in this direction. The American chemist Theodore William Richards analyzed compounds with fantastic precautions against impurities and error and obtained such accurate atomic weight values that he received the 1914 Nobel Prize in chemistry for his work.

But as fate would have it, by that time atomic weights had gotten away from chemists and entered the domain of the physicist.

The break came with the discovery of the subatomic particles in the 1890s. The atom was *not* a featureless spherical particle, it turned out. It was a conglomerate of still smaller particles, some of which were electrically charged.

It turned out then that the fundamental distinction between atoms of different elements was not the atomic weight at all but the quantity of positive electric charge upon the nucleus of the atom. (Again, this fitted the intellectual fashion of the times, for as the nineteenth century wore on, the mechanical Newtonian universe gave way to a universe of force fields according to the theories of the English chemist Michael Faraday and the Scottish physicist James Clerk Maxwell. Electric charge fits into this force field scheme.)

It turned out that most elements consisted of varieties of atoms of somewhat different atomic weight. These varieties are called "isotopes."

What we have been calling the atomic weight is only the average of the weights of the various isotopes making up the element.

Physicists began to determine the relative masses of the individual isotopes by nonchemical methods, with a degree of accuracy far beyond the ordinary chemical methods even of Nobel laureate Richards. To get an accurate atomic weight it was then only necessary to take a weighted average of the masses of the isotopes making up the elements, allowing for the natural percentage of each isotope in the element as found in nature.

The fact that atomic weights had thus become a physical rather than a chemical measurement might not have been embarrassing, even for the most sensitive chemist, were it not for the fact that physicists began to use atomic weight values slightly different from that used by the chemists. And what made it really bad was that the physicists were right and the chemists wrong.

Let me explain.

MICHAEL FARADAY

Faraday, born in Newington, Surrey, on September 22, 1791, was one of the ten children of a blacksmith who moved, with his brood, to London. There was no money to educate young Faraday past reading and writing and he was apprenticed to a bookbinder in 1805.

This, as it happened, was a stroke of luck, for he was exposed to books and dipped into every book he bound. His employer, for a wonder, was sympathetic and not only allowed him to read but gave him time off to attend scientific lectures.

In 1812 a customer gave Faraday tickets to attend the lectures of Humphry Davy, then England's most famous scientist. Faraday took careful notes which he further elaborated with colored diagrams. He ended with 386 pages which he bound in leather and sent to Davy, who offered him a job as bottle washer at a salary smaller than he had earned as a bookbinder.

Almost at once Davy left for a trip to Europe, taking Faraday along as valet and servant.

Back in England eventually, Faraday worked industriously in Davy's laboratory and gradually showed ever-clearer signs of genius. He developed methods for liquefying gas, discovered benzene, and worked on the manner in which electric current passed through solutions. He worked out what are still called "Faraday's laws of electrolysis."

He studied the interrelation of electricity and magnetism and worked out the first electric generator—the first practical devices for turning heat and mechanical energy into a continuous flow of electric current, on which rests the entire electrified world of today.

Unfortunately, the young bottle washer who grew to be far greater than his master incurred the bitter envy of that master and Davy did what he could to interfere with Faraday's career. Faraday, a deeply religious man, never allowed himself to be betrayed into anger over this. Simple and unassuming always, he died near London on August 25, 1867, universally accepted as one of the greatest scientists in history.

From the very beginning, the measurement of atomic weights had required the establishment of a standard. The most logical standard seemed to be that of setting the atomic weight of hydrogen equal to 1. It was suspected then (and it is known now) that hydrogen possessed the lightest possible atom, so setting it equal to 1 was the most natural thing in the world.

The trouble was that in determining atomic weights one had to start with equivalent weights. (In the beginning, anyway.) To determine equivalent weights, one needed to work with two elements that combined easily. Now hydrogen combined directly with but few ele-

THEODORE WILLIAM RICHARDS

Richards was born in Germantown, Pennsylvania, on January 31, 1868. His father was a painter and his mother a poet and he himself inherited talents in both directions. In addition, he was interested in music and, of course, in science.

For his doctor's degree, which he earned in 1888 at Harvard, he undertook to determine a more accurate value for the ratio of the atomic weight of oxygen to that of hydrogen. He continued this line of work after he had earned his degree and dedicated his professional life to determining with the greatest possible accuracy the atomic weights of the various elements.

Over nearly three decades he and his students established the atomic weight of some sixty elements with an accuracy that seemed to represent the limit of what could be done with purely chemical methods. For this he received the Nobel Prize for chemistry in 1914.

His work brought the age of classical atomic weight determinations to an end and marked the initiation of a new age as well. In 1913 he began the determination of the atomic weights of lead from different minerals and detected small but definite differences. This provided experimental verification of the fact that elements could be mixtures of different isotopes each with its own characteristic atomic weight—and these mixtures could vary from sample to sample.

The existence of isotopes, to which Richards had thus contributed evidence, showed that ordinary atomic weights, though still a matter of importance for chemical calculations, were no longer fundamental physical data.

Richards died in Cambridge, Massachusetts, on April 2, 1928.

ments, whereas oxygen combined directly with many. It was a matter of practical convenience to use oxygen, rather than hydrogen, as a standard.

This made a slight modification necessary.

Atomic weights, after all, don't match in exact whole-number ratios. If the atomic weight of hydrogen is set at exactly 1, then the atomic weight of oxygen is not quite 16. It is, instead, closer to 15.9. But if oxygen is the element most often used in calculating equivalent weights, it would be inconvenient to be forever using a figure like 15.9. It is an easy alternative to set the atomic weight of oxygen exactly at

16 and let the atomic weight of hydrogen come out a trifle over 1. It comes out to 1.008, in fact.

We can call this the "O=16" standard. It made chemists very happy, and there arose nothing to challenge it until the 1920s. Then came trouble.

Oxygen, it was discovered in 1929, was a mixture of three different isotopes. Out of every 100,000 oxygen atoms, 99,759, to be sure, had an atomic weight of about 16. Another 204, however, had an atomic weight of about 18, while the remaining 37 had an atomic weight of 17. (The isotopes can be symbolized as O^{16}, O^{17} and O^{18}.)

This meant that when chemists set oxygen equal to 16, they were setting a weighted average of the three isotopes equal to 16. The common oxygen isotope was just a little under 16 (15.9956, to be exact), and the masses of the relatively few oxygen atoms of the heavier isotopes pulled that figure up to the 16 mark.

Physicists working with individual nuclei were more interested in a particular isotope than in the arbitrary collection of isotopes in an element. In this they had logic on their side, for the mass of an individual isotope is, as far as we know, absolutely constant, while the average mass of the atoms of an element fluctuates slightly as the mixture varies a tiny bit under different conditions.

Now we have two scales. First there is the "chemical atomic weight" on the "O=16" standard. Second, there is the "physical atomic weight" on the "O^{16}=16" standard.

On the chemical atomic weight scale, the atomic weight of oxygen is 16.0000; while on the physical atomic weight scale, the heavier oxygen isotopes pull the average weight up to 16.0044. Naturally, all the other atomic weights must change in proportion, and every element has an atomic weight that is 0.027 per cent higher on the physical scale than on the chemical scale. Thus, hydrogen has a chemical atomic weight of 1.0080 and a physical atomic weight of 1.0083.

This isn't much of a difference, but it isn't neat. Chemists and physicists shouldn't disagree like that. Yet chemists, despite the weight of logic against them, were reluctant to abandon their old figures and introduce confusion when so many reams of calculations in the chemical literature had been based on the old chemical atomic weights.

Fortunately, after three decades of disagreement, a successful compromise was reached.

It occurred to the physicists that, in using an "$O^{16}=16$" standard, they were kowtowing to a chemical prejudice which no longer had validity. The only reason that oxygen was used as the standard was the ease with which oxygen could be used in determining equivalent weights.

But the physicists weren't using equivalent weights; they didn't give a continental for equivalent weights. They were determining the masses of charged atoms by sending them through a magnetic field of known strength and measuring the effect upon the paths of those atoms.

In this connection, oxygen atoms were not the best atoms to use as standard; carbon atoms were. The mass of the most common carbon isotope, C^{12}, was more accurately known than that of any other isotope. Moreover, C^{12} had a mass that was 12.003803 on the physical scale and was almost exactly 12 on the chemical scale.

Why not, then, set up a "$C^{12}=12$" scale? It would be just as logical as the "$O^{16}=16$" scale. What's more, the "$C^{12}=12$" scale would be almost exactly like that of the chemical "$O=16$" scale.

In 1961, the International Union of Pure and Applied Physics issued a ukase that this be done. The mass of C^{12} was set at exactly 12.000000, a decrease of 0.033 per cent. Naturally, the masses of all other isotopes had to decrease by exactly the same percentage, and the "$C^{12}=12$" scale fell very slightly below the "$O=16$" scale.

Thus, the chemical atomic weights of hydrogen and oxygen were 1.0080 and 16.0000 respectively. The atomic weights on the new physical scale were 1.00797 and 15.9994 respectively.

The difference is now only 0.003 per cent, only one-tenth of the difference between the chemical scale and the old physical scale.

The chemists could no longer resist; the difference was so small that it would not affect any of the calculations in the literature. Consequently, the International Union of Pure and Applied Chemistry has also accepted the "$C^{12}=12$" scale. Physicists and chemists once again, as of 1961, speak the same atomic weight language.

Note, too, how it was done. The physicists made most of the adjustment in actual value and that was a victory for the chemists. The chemists, on the other hand, adopted the logic of the single isotope as standard, and that was a victory for the physicists.

And since the standard which has now been adopted is the most accurate one yet, the net result is a victory for everybody.

Now *that* is the way to run the world—but I'll refrain from trying to point a moral.

2

SLOW BURN

For many years now I have been an inveterate admirer of Sir Isaac
Newton. One can, after all, make out a good case for his having been
the greatest scientist who ever lived.

What's more, it doesn't displease me one little bit that Newton's
first name is Isaac. To be sure, I wasn't named for him, but for my
grandfather. Yet the principle remains; we have something in common. And to top it off, the Boston suburb in which I live[1] is named
Newton—how do you like that?

So you see, I have lots of reasons for being an Isaac Newton fan
and it therefore pains me to admit there are flaws in the shining picture
he presents. In physics and astronomy he was a transcendent genius.
In mathematics he was a ground-breaking prodigy. Yet in chemistry he
was nothing but a bumbler. He wasted his time in a vain and useless
effort to manufacture gold, scouring Europe for recipes, trying each
one and forever being disappointed.

[1] I moved to New York eight years after this article was written. I had lived in
Newton for fourteen years.

This is a dramatic way of showing that Newton stood at a midway point in the history of the physical sciences. In the 1680s when he announced his laws of motion and his theory of gravitation, the birth of modern physics (thanks to Galileo) was just one century in the past and the birth of modern chemistry (thanks to Lavoisier) was just one century in the future.

Now the story of the birth of physics has been told and told again. We all know (or should) about Galileo's experiments with falling bodies which, at one stroke, destroyed Aristotelian physics and established the modern form of the science. In popular mythology this is concentrated into a single experiment, the dropping of a heavy and light ball from the top of the Leaning Tower of Pisa and watching them hit the ground simultaneously. (Actually, it is quite certain that Galileo never performed this experiment.)

On the other hand, the birth of chemistry is graced by no such key experiment. There is no chemical equivalent of dropping weights off the Leaning Tower of Pisa; no single, classic feat to go ringing down the corridors of time as the smasher of the old and the beginner of the new. At least, I don't find one in the books I've read on the subject; not one that is pointed to as *the* experiment.

Except that I think I've found one. I think I can make a case for the existence of a single, simple experiment that smashed the old chemistry and started the new chemistry. It was every bit as dramatic and conclusive (if not quite as spectacular) as the Leaning Tower of Pisa experiment, except that:

(1) The crucial chemical experiment really happened and is not a myth, and

(2) It involved a mad scientist and should therefore strike a nostalgic chord in the hearts of all true science-fiction fans.

With your permission then, O Gentle Reader (or, if necessary, without), I shall tell the story of the birth of Modern Chemistry, as I see it.

In the time of Newton chemical theory was still based, in large part, on what the Greek philosophers had worked out two thousand years earlier. The "four elements" (that is, the fundamental substances out of which the universe was made) were earth, water, air, and fire.

The Greek philosophers felt that actual bodies were made up of the four elements in particular proportions. One could well imagine, then, that the elements in one body could be separated and then recombined in different proportions to form a second body of a different sort. In this way, one could change one metal into another (if one could but discover the correct procedure), and in particular, one could change lead into gold.

For about fifteen hundred years, alchemists tried to find out the proper recipe for such "transmutation." The Arabs, in the process, worked out the theory that there were two special principles involved in the different solid bodies with which they worked. There was the metallic principle, mercury, and the combustible principle, sulfur.

This didn't help them make gold, and by Newton's time chemistry seemed badly in need of some new ideas. What's more, any new ideas that did come along ought to deal with combustion. Coal was beginning to come into use as a new fuel. Men were beginning to play with the steam produced by the heat of burning fuel. In general, the matter of combustion was in the air and as exciting in 1700 as electricity was to be in 1800, radioactivity in 1900, and rocketry in 1950.

Onto the scene then, steps a German physician named Georg Ernest Stahl. While still in his twenties he was appointed court physician to the Duke of Weimar. In later life he was to become physician to still higher royalty, King Frederick William I of Prussia. His lectures on medicine at the University of Halle were famous and well attended.

In 1700 this man advanced a theory of combustion that made more sense than anything previously suggested. He drew heavily on alchemical notions and, in particular, on the combustible principle, sulfur. He gave this principle a new name and described its behavior in greater detail.

The principle he called "phlogiston," from a Greek word meaning "to set on fire," for he held that all inflammable objects contained phlogiston and it was only the phlogiston content that made it possible for them to burn.

During the process of combustion, said Stahl, the burning material lost its content of phlogiston, which poured out into and was received by the air. What was left after combustion was completely lacking in phlogiston and could burn no more. Wood and coal, for instance, were rich in phlogiston, but the ash they left behind contained none.

Stahl's greatest contribution to chemical thinking was his suggestion that the process of rusting of metals was similar in principle to that of the burning of wood. A metal, such as iron, was rich in phlogiston. As it corroded, it lost phlogiston to the air, and when all the phlogiston was gone, only rust was left behind.

The basic difference, then, between the burning of wood and the rusting of iron was no more than a matter of speed. Wood lost phlogiston so rapidly that the velocity of its passing made it visible as flame. Iron lost phlogiston so slowly that its passage was imperceptible. Burning, in Stahl's view, was a fast rusting, while rusting was a slow burn.

In this, Stahl was quite correct, but he gets little credit for it. About the first thing chemistry students are taught to do is to laugh at the phlogiston theory, so that Stahl is either forgotten or condemned, and I consider that unfair.

As a matter of fact, the phlogiston theory explained quite a few things that were not explained before, most notably the matter of metallurgy. For instance, it had been known for thousands of years that if metal ore were heated strongly, in contact with burning wood or charcoal, the free metal could be obtained. As for *why* this happened, no one had a good answer.

Until Stahl, that is. According to the phlogiston theory, it was easy to see that a metal ore was a kind of naturally occurring rust that was completely free of phlogiston and therefore showed no metallic properties. If heated in the presence of phlogiston-rich charcoal, phlogiston passed from the charcoal to the ore. As the ore gained phlogiston, it turned into metal. As the charcoal lost phlogiston, it turned into ash.

Isn't that neat?

Unfortunately, there was one great flaw in the theory. When a metal rusted, it gained weight! One pound of iron produced about one and a half pounds of iron rust. If the conversion were the result of the loss of phlogiston and not the gaining of anything, where did the extra weight come from?

A few chemists worried about this and tried to explain that phlogiston had negative weight! Instead of phlogiston being pulled down by gravity, it was pushed up by levity. (You may take that as a pun, if you choose, but levity was the actual term used.) Thus, a pound of iron could be considered as containing minus half a pound of phlogiston, and when the phlogiston left, the resulting rust would weigh one and a half pounds.

This notion went over like a lead balloon. For one thing, no example of levity was found anywhere in nature outside of phlogiston, and for another, when wood burned it *lost* weight. The ash it left behind was much lighter than the original wood. If the wood had lost phlogiston and if phlogiston exerted a force of levity, why wasn't the ash heavier than the wood, as rust is heavier than iron?

There was no answer to this, and the average chemist of the day simply shrugged. There was, after all, no tradition of exact measurement in chemistry. For thousands of years everyone had worked the chemical industries as art forms rather than as sciences. The alchemists had involved themselves in purely descriptive observations. They had noted the formation of precipitates, the emission of vapors, the changes of colors—but such things as weight and volume were irrelevant.

For two generations matters continued thus, and then, in the 1770s, a number of momentous developments took place. For one thing, chemists began to concern themselves with air.

To the ancient Greeks air was an element, a single substance. However, the Scottish chemist Joseph Black burned a candle in a closed container of air, as the 1770s opened, and found that the candle eventually went out. When it did, there was still plenty of air in the container, so *why* did it go out?

He was busy with other matters, so he passed the problem on to a student of his named Daniel Rutherford. (Rutherford, by the way, was the uncle of the poet and novelist Sir Walter Scott.)

In 1772 Rutherford repeated Black's experiments and went further. New candles, set on fire and placed in the air remaining after the old candle had burned out, promptly went out themselves. Mice, placed in such air, died.

Rutherford analyzed these observations in terms of the phlogiston theory. When a candle burned in an enclosed volume of air, it gave up phlogiston to the air but, apparently, any given volume of air could only hold so much phlogiston and no more. When the air was filled with phlogiston, the candle went out and nothing further would burn in that air. A living creature which, in the process of breathing, constantly gave up phlogiston (there had been speculations dating back to Roman times that respiration was analogous to combustion) could not do so in this phlogiston-filled air, and died. Rutherford called this asphyxiating gas "phlogisticated air."

The scene now shifts southward to England, where a Unitarian minister Joseph Priestley had become interested in science after he met the American scientist and statesman Benjamin Franklin in 1766.

Priestley's great discovery came from experiments with mercury in 1774. He began by heating mercury with sunlight concentrated through a large magnifying glass. The heat caused the gleaming surface of the mercury to be coated with a reddish powder. Priestly skimmed off the

powder and heated it to a still higher temperature. The powder evaporated, forming two different gases. One of these was mercury vapor, for
it condensed into droplets of mercury in the cool upper regions of the
vessel. The other remained an invisible vapor.

How did Priestley know it was there? Well, it had peculiar properties
that were not like those of ordinary air. A smoldering splint of wood
thrust into the container in which the red powder from mercury was being heated burst into bright flame. Priestley collected the vapors and
found a candle would burn in it with unearthly brightness; he found
that mice placed in the vapor would jump about actively; he even
breathed some himself and reported it made him feel very "light and
easy."

Priestley interpreted all this according to the phlogiston theory.
When mercury was heated it lost some of its phlogiston to air and became a red powder which lacked phlogiston and could be considered a
kind of mercury rust. If he heated this mercury rust strongly, it absorbed
phlogiston from the air and became mercury again. Meanwhile, the air
in the neighborhood was bled of its phlogiston and became "dephlogisticated air." Naturally, such dephlogisticated air was unusually thirsty for
phlogiston. It sucked phlogiston rapidly out of a smoldering splint and
the velocity of the reaction was visible as a burst of flame. For similar
reasons, candles burned more brightly and mice ran about more actively
in dephlogisticated air than in ordinary air.

The Priestley and Rutherford experiments, taken together, seemed to
show that air was a single material substance, which could be altered in
properties by a variation in its content of the imponderable fluid,
phlogiston.

Ordinary air contains some phlogiston but is not saturated with it. It
can gain phlogiston when something burns in it; or it can lose phlogiston when a rust heated in it becomes a metal. When it gains all the
phlogiston it can hold, it will no longer support combustion or life and
it is then Rutherford's gas. If it loses all the phlogiston it has, it will
support combustion with great eagerness and life with great ease and
will then be Priestley's gas.

Now we shift still farther south. In Paris the brilliant young chemist
Lavoisier is working hard under the stress of an idea—that measurement is as important to chemistry as Galileo showed it to be to physics. Qualitative observations are insufficient; one must be quantitative.

As an example—when water, even the purest, was slowly boiled away in a glass vessel, some sediment was always left behind. Alchemists had often done this and they had pointed to the sediment as an example of the manner in which the element water had been converted to the element earth. (From this they deduced that transmutation was possible and that lead could be turned to gold.)

About 1770 Lavoisier decided to repeat the experiment, but quantitatively. He began by accurately weighing a clean flask and adding an accurate weight of water. He then boiled the water under conditions

JOSEPH PRIESTLEY

Priestley was born in Fieldhead, Yorkshire, on March 13, 1733. His mother died when he was seven and he was brought up by a pious aunt. He was an exceedingly bright boy who sopped up knowledge almost indiscriminately, learning a variety of languages, including Hebrew and Arabic, and teaching himself science.

He was the son of a Nonconformist preacher and turned to the left both in religion and politics, becoming a Unitarian preacher and openly supporting the American colonists in their war against George III.

In 1766, on one of his periodic visits to London, Priestley met Benjamin Franklin and became interested in electricity. Priestley was the first to find that carbon was a conductor of electricity and wrote an important history of electrical research.

He discovered that fermenting grain produced a gas which turned out to be carbon dioxide. He found that a solution of the gas produced a pleasant drink ("seltzer") and flavored seltzer is what we now drink as soda pop.

He collected gases by passing them through mercury instead of through water so that he was the first to study water-soluble gases such as ammonia and hydrogen chloride. In 1774 he heated mercuric oxide and discovered the gas that later came to be called "oxygen," noting the odd way in which it encouraged burning. (This was an important landmark in the history of chemistry.)

His sympathy with the French Revolutionists, added to his earlier sympathy with the Americans, made him a marked man. In 1791 a mob burned down his house in Birmingham and he had to escape to London. In 1794 he left for the United States (which was now independent) and there he spent the last decade of his life in peace, gaining the friendship of Thomas Jefferson.

He died in Northumberland, Pennsylvania, on February 6, 1804.

so designed that the rising water vapor was cooled, condensed back to water, and forced to drip again into the still-boiling contents of the flask. He continued this for 101 days, thus giving the water plenty of time to turn into earth. He then stopped and let all the water cool down.

Sure enough, as the water cooled, the sediment formed. Lavoisier poured out the water, filtered off the sediment, and weighed each separately. The weight of the water had not changed at all. He then weighed the flask. The flask had lost weight and the loss in weight was just equal to the weight of the sediment. Water had not changed to earth; it had simply dissolved some of the material of the flask.

Thus he showed that one conclusion drawn from a particular experiment could be shifted to another and much more plausible conclusion by simply becoming quantitative.

In a later experiment Lavoisier put some tin in a vessel which he then closed. He next weighed the whole business accurately. Then he heated the vessel.

A white rust formed on the tin. It was known that such a rust was invariably heavier than the original metal, yet when Lavoisier weighed

ANTOINE LAURENT LAVOISIER

Lavoisier was born in Paris, on August 26, 1743. Loved and pampered, first by his mother and then, after her early death, by an adoring aunt, he was given an excellent education and proved a brilliant student. His father, a lawyer, hoped his son would follow in that profession, but the young man veered toward science.

From the very beginning of his chemical researches, Lavoisier recognized the importance of accurate measurement. It was by careful weighings that he showed mass did not change in the course of chemical reactions (at least not to an extent measurable in those days), thus establishing the law of conservation of mass.

He showed that rusting and burning was accompanied by the absorption of part of the air but not all, and he demonstrated that the air was a mixture of an active gas he called "oxygen" and an inactive one which we now call "nitrogen," the latter being incapable of supporting combustion or life, while the former could.

He wrote the first modern textbook in chemistry and, with others, worked out the system of nomenclature that chemists still use today —so that he is rightly called the Father of Modern Chemistry.

However, Lavoisier invested money in a private tax-collecting firm that was hated by the taxpayers and married the daughter of an important executive of the firm. He plowed his profits back into his

researches but it was a deadly connection, especially since he offended a certain Jean-Paul Marat, a journalist who fancied himself a scientist and had those pretensions pricked by Lavoisier.

Once the French Revolution broke out, Marat became a powerful revolutionary leader and, eager for revenge, denounced Lavoisier regularly. Lavoisier's connection with the tax collectors offered an easy weapon with which to attack him. He was arrested by someone who is supposed to have said, "The Republic has no need of scientists," and was guillotined in Paris on May 8, 1794. The radicals were overthrown two months later, but Lavoisier's head could not be restored.

the whole setup, he found the total weight had not changed at all. If the rust were heavier than the tin, then that gain in weight must have been countered by an equal loss in weight elsewhere in the vessel. If the loss in weight were in the air content, then a partial vacuum should now exist in the vessel. Sure enough, when Lavoisier opened the vessel, air rushed in and then the system increased in weight. The increase was equal to the extra weight of the rust.

Lavoisier therefore suggested the following: Combustion (or rust-formation) was caused not by the loss of phlogiston but by the combination of the fuel or metal with air. Phlogiston had nothing to do with it. Phlogiston did not exist.

The weak point in this new suggestion, just at first, lay in the fact that not all the air was involved in this. Lavoisier found that when a candle burned, it used up only about one fifth of the air. It would burn no longer in the remaining four fifths.

Light dawned when Priestley visited France and had a conversation with Lavoisier. Of course! Lavoisier rushed back to his work. If phlogiston did not exist, then air could not change its properties with gain or loss of phlogiston. If two kinds of air seemed to exist with different properties, then it was because air contained two different substances.

The one fifth of the air which a burning candle used up was Priestley's dephlogisticated air, which Lavoisier now called "oxygen," from Greek words meaning "sourness-producer." (Lavoisier thought oxygen was a necessary component of acids. It isn't, but the name will never be changed now.) As for the remaining four fifths of the air, that portion in which candles would not burn and mice would not live, that was Rutherford's phlogisticated air, and Lavoisier called it "azote," from Greek words meaning "no life." Nowadays, we call it "nitrogen."

Air, according to Lavoisier, then, was one fifth oxygen and four fifths nitrogen. Combustion and rusting were brought about by the combination of materials with oxygen only. Some combinations (or "oxides"), such as carbon dioxide, were vapors and left the scene of combustion altogether, which was why coal, wood, and candles all lost weight drastically after burning. Other oxides were solids and remained on the spot, which was why rust was heavier than metal—heavier by the added oxygen.

In order for a new theory to displace an old comfortable one, the new theory has to be *clearly* better, and the oxygen theory was not, just at first. To most chemists, oxygen just seemed phlogiston in reverse.

Instead of wood losing phlogiston in combustion, it gained oxygen. Instead of iron ore gaining phlogiston in iron smelting, it lost oxygen.

Lavoisier could only have carried conviction by proving that the matter of weight was crucial, for the oxygen theory explained the weight changes in combustion and rusting, while the phlogiston theory did not and could not.

Lavoisier tried to emphasize the importance of weight and to make it central to chemistry by maintaining that there was no change in total weight during the course of any chemical reaction in a *closed* system, where vapors were not allowed to escape and outside air could not be added. This is the "law of conservation of mass." Another way of putting it is that matter can neither be created nor destroyed, and if that is true, then the phlogiston theory is fallacious, for in it the added weight of the rust appears out of nowhere and matter must therefore be created.

Unfortunately, Lavoisier could not make the law of conservation of mass hard and fast at first. There was a flaw. Lavoisier tried to measure the amount of oxygen a human being absorbed in breathing and to compare it with the carbon dioxide he exhaled. When he did that, it always turned out that some of the oxygen had disappeared. The exhaled carbon dioxide never accounted for all the oxygen taken in. If the law of conservation of mass didn't hold, there was no handy stick with which to kill phlogiston.

Now let's go back to England and to our mad scientist, Henry Cavendish.

Cavendish, you see, was pathologically shy and unbelievably absent-minded. It was only with difficulty that he could speak to one man; and it was virtually impossible to speak to more than one. Although he regularly attended dinner at the Royal Society, dressed in snuffy, old-fashioned clothes, he ate in dead silence with his eyes on his plate.

He was a woman-hater (or, perhaps, woman-fearer) to the point where he could not bear to look at one. He communicated with his female servants by notes, and any who accidentally crossed his path in his house was fired on the spot. He built a separate entrance to his house so he could come and leave alone. In the end, he even insisted on dying alone.

He came of a noble family and at the age of forty inherited a fortune, but paid no particular attention to it. Money literally meant nothing to him, and neither did fame. Many of his important discoveries he never bothered publishing, and it is only by going through the papers he left behind that we know of them.

Some discoveries, however, he did publish. In 1766, for instance, he discovered an inflammable gas produced by the action of acids on metals. This had been done before, but Cavendish was the first to study the gas systematically and so he gets credit for its discovery.

One thing that Cavendish noted about the gas was that it was exceedingly light—far lighter than air; lighter than any material object then known (or since discovered). With his mind on the "levity" that some had suggested as one of the properties of phlogiston, Cavendish began to wonder whether he had stumbled on something that was mostly, or even entirely, phlogiston. Perhaps he had phlogiston itself.

After all, as the gas left the metal through the action of acids, the metal formed a rust with phenomenal rapidity. Furthermore, the gas was highly inflammable; indeed, explosively so; and surely that was to be expected of phlogiston.

When, in the decade that followed, Rutherford isolated his phlogisticated air and Priestley his dephlogisticated air, it occurred to Cavendish that he could perform a crucial experiment.

He could add his phlogiston to a sample of dephlogisticated air and convert it first into ordinary air and then into phlogisticated air. If he did that, it would be ample proof that his inflammable gas was indeed phlogiston and, moreover, it would be a general proof of the truth of the phlogiston theory.

So, in 1781, Cavendish performed *the* crucial experiment in chemistry. It was simplicity itself. He merely set acid to working on metal so that a jet of his phlogiston could be forced out of a glass tube. This jet of phlogiston could be lighted by a spark and allowed to burn inside a vessel full of dephlogisticated air. That was all there was to it.

But when he did it, he found to his surprise that he had not formed phlogisticated air at all. Instead, the inner walls of the vessel were bedewed with drops of a liquid that looked like water, tasted like water, felt like water, had all the chemical properties of water and, egad, sir, *was* water.

Cavendish hadn't proved the phlogiston theory at all. In fact, as Lavoisier saw at once, Cavendish's experiment had once and for all killed phlogiston.

As soon as Lavoisier heard of Cavendish's work, he jumped upon it with loud cries of delight. He repeated the experiment with improvements and named Cavendish's gas "hydrogen," from Greek words meaning "water-producer," a name it keeps to this day.

Here's what this one simple experiment of Cavendish's did:

(1) It proved water to be an oxide; the oxide of hydrogen. This was the last, final blow to the "four-elements" theory of the Greeks, for water was not a basic substance after all.

(2) It wiped out the notion that air was a single substance varying in properties according to its phlogiston content. If that were so, then hydrogen plus oxygen would yield nitrogen (as Cavendish had, in truth, expected it would—using the eighteenth-century terminology of phlogisticated air, dephlogisticated air, and so on). But if air were not a single substance, then the only way of accounting for the experiments of the 1770s was to assume it a mixture of two substances.

(3) Lavoisier realized that the foodstuffs that underwent combustion in the body contained both carbon *and* hydrogen. In the light of Cavendish's experiment, then, it was not surprising that the carbon dioxide produced by the body was less than sufficient to account for the oxygen absorbed. Some of the oxygen was used up in combining with hydrogen to form water, and expired breath was rich in water as well as in carbon dioxide. The obvious flaw in the law of conservation of mass was removed. The importance of quantitative measurement in chemistry was thus established and has never since been doubted.

In short, then, all of Modern Chemistry traces back, clean and true as an arrow, to Cavendish's burning jet of hydrogen.

There is an ironic postscript to the story, though. Lavoisier had one flaw in an otherwise admirable character. He had a tendency to grab for credit that did not belong to him. In advancing his theory of combustion, for instance, he never mentioned Priestley's experiments and never indicated that he had discussed them with Priestley. In fact, he tried to give the impression that he, himself, was the discoverer of oxygen. In the same way, when he repeated Cavendish's experiment of burning hydrogen, he tried to give the impression, without quite saying so, that the experiment was original with him.

Lavoisier didn't get away with these little tricks and posterity has forgiven him his vanity, for what he *did* do (including a deal of material I haven't mentioned in this article) was quite enough for a hundred ordinary chemists.

However, it is quite likely that neither Priestley nor Cavendish felt particularly kindly toward Lavoisier as a result. At least, neither man would accept Lavoisier's new chemistry. Both men refused to abandon phlogiston, and remained stubborn devotees of the old chemistry to the end of their lives.

Which once again proves, I suppose, that scientists are human. Like the metals they work with, they can be subjected to the effects of a slow burn.

3

THE ELEMENT OF PERFECTION

In the old days of science fiction, when writers had much more of the leeway that arises out of scientific innocence, a "new element" could always be counted on to get a story going or save it from disaster. A new element could block off gravity, or magnify atoms to visible size, or transport matter.

Much of this "new element" fetish was the outcome of the Curies' dramatic discovery in uranium ore of that unusual element, radium, in 1898. And yet, that same decade, another element was found in uranium ore under even more dramatic circumstances. Though this second element aroused nothing like the furore created by radium, it proved, in the end, to be the most unusual element of all and to have properties as wild as any a science-fictioneer ever dreamed up.

Furthermore, the significance of this element to man has expanded remarkably in the past five years,[1] and thoughts connected with that expansion have brought a remarkable vision to my mind which I will describe eventually.

[1] *And has remained expanded in the fourteen years since this article appeared.*

In 1868, there was a total eclipse of the sun visible in India, and astronomers assembled jubilantly in order to bring to bear a new instrument in their quest for knowledge.

This was the spectroscope, developed in the late 1850s by the German scientists Gustav Robert Kirchhoff and Robert Wilhelm Bunsen. Essentially this involved the conduction of the light emitted by heated elements through a prism to produce a spectrum in which the wave lengths of the light could be measured. Each element produced light of wave lengths characteristic of itself, so that the elements were "fingerprinted."

The worth of this new analytical method was spectacularly demonstrated in 1860 when Kirchhoff and Bunsen heated certain ores, came across spectral lines that did not jibe with those already known and, as a result, discovered a rare element, cesium. The next year, they showed this was no accident by discovering another element, rubidium.

With that record of accomplishment, astronomers were eager to turn the instrument on the solar atmosphere (unmasked only during eclipses) in order to determine its chemical composition across the gulfs of space.

Almost at once, the French astronomer Pierre J. C. Janssen observed a yellow line that did not quite match any known line. The English astronomer Norman Lockyer, particularly interested in spectroscopy, decided this represented a new element. He named it for the Greek god of the sun, Helios, so that the new element became "helium."

So far, so good, except that very few, if any, of the earthly chemists cared to believe in a nonearthly element on the basis of a simple line of light. Lockyer's suggestion was greeted with reactions that ran the gamut from indifference to mockery.

Of course, such conservatism appears shameful in hindsight. Actually, though, hindsight also proves the skepticism to have been justified. A new spectral line does not necessarily signify a new element.

Spurred on by the eventual success of helium, other "new elements" were found in outer space. Strange lines in the spectra of certain nebulae were attributed to an element called "nebulium." Unknown lines in the sun's corona were attributed to "coronium" and similar lines in the auroral glow to "geocoronium."

These new elements, however, proved to be delusions. They were produced by old, well-known elements under strange conditions duplicated in the laboratory only years afterward. "Nebulium" and "geocoronium" turned out to be merely oxygen-nitrogen mixtures under highly ionized conditions. "Coronium" lines were produced by highly ionized metals such as calcium.

So you see that the mere existence of the "helium" line did not really prove the existence of a new element. However, to carry the story on, it is necessary to backtrack still another century to a man even further ahead of his times than Lockyer was.

In 1785, the English physicist Henry Cavendish was studying air, which at the time had just been discovered to consist of two gases, oxygen and nitrogen. Nitrogen was an inert gas; that is, it would not combine readily with other substances, as oxygen would. In fact, nitrogen was remarkable for the number of negative properties it had. It was colorless, odorless, tasteless, insoluble, and incombustible. It was not poisonous in itself, but neither would it support life.

Cavendish found that by using electric sparks, he could persuade the nitrogen to combine with oxygen. He could then absorb the resulting compound, nitrogen oxide, in appropriate chemicals. By adding more oxygen he could consume more and more of the nitrogen until finally his entire supply was reduced to a tiny bubble which was about 1 per cent of the original volume of air. This last bubble he could do nothing with, and he stated that in his opinion there was a small quantity of an unknown gas in the atmosphere which was even more inert than nitrogen.

Here was a clear-cut experiment by a first-rank scientist who reached a logical conclusion that represented, we now realize, the pure truth. Nevertheless, Cavendish's work was ignored for a century.

Then, in 1882, the British physicist John William Strutt (more commonly known as Lord Rayleigh, because he happened to be a baron as well as a scientist) was investigating the densities of hydrogen and oxygen gas in order better to determine their atomic weights, and he threw in nitrogen for good measure. To do the job with the proper thoroughness, he prepared each element by several different methods. In the case of hydrogen and oxygen, he got the same densities regardless of the method of preparation. Not so in the case of nitrogen.

He prepared nitrogen from ammonia and obtained a density of 1.251 grams per liter. He also prepared nitrogen from air by removing the oxygen, carbon dioxide, and water vapor, and for that nitrogen he obtained a density of 1.257 grams per liter. This discrepancy survived his most careful efforts. Helplessly, he published these results in a scientific journal and invited suggestions from the readers, but none were received. Lord Rayleigh, himself, thought of several possible explanations: that the atmospheric nitrogen was contaminated with the heavier oxygen, or with the triatomic molecule N_3, a kind of nitrogen analog of ozone; or that the nitrogen from the ammonia was contaminated with the lighter hydrogen or with atomic nitrogen. He checked out each possibility and all failed.

Then, a decade later, a Scottish chemist, William Ramsay, came to work for Lord Rayleigh and, tackling the nitrogen problem, harked back to Cavendish and wondered if the atmosphere might not contain small quantities of a gas that remained with nitrogen when everything else was removed and which, being heavier than nitrogen, gave atmospheric nitrogen a spuriously high density.

In 1894, Ramsay repeated Cavendish's original experiment with improvements. He passed atmospheric nitrogen over red-hot metallic magnesium. Nitrogen wasn't so inert that it could resist that. It reacted

HENRY CAVENDISH

Cavendish, an Englishman, was born at Nice, France, on October 10, 1731. He was born there because his mother was on a trip to improve her health in the warmth of the Riviera. She died when her son was two years old, however.

Cavendish spent four years at Cambridge but never took his degree, partly because he could not face the professors during the necessary examinations. All his life he had difficulty facing people— especially women, whom he feared with psychotic intensity.

He had one and only one love and that was scientific research. He spent almost sixty years on it and did not even care whether his findings were published or not, so long as his own curiosity was sated. As a result, much of what he did remained unknown until years after his death. His experiments on electricity in the early 1770s anticipated most of what was to be discovered in the next half century, yet it remained unpublished, and there is no way of estimating what that unnecessary secrecy cost the human race in scientific progress. He had no talent for inventing instruments and measured the strength of a current by estimating the pain produced when he used it to shock himself.

He was the first to study hydrogen and to show that it formed water on burning. He removed the oxygen and nitrogen from air and showed that a small bubble of gas remained, one that was eventually identified as argon. Most spectacularly of all, he measured the strength of the gravitational constant in laboratory experiments and was the first man to calculate the mass of the Earth.

He suffered few economic pressures. He came of a noble family and at the age of forty inherited a fortune of over a million pounds but paid no particular attention to it. On his death, the fortune, despised and virtually untouched, went to relatives. He died in London, on February 24, 1810—dying as he lived, entirely alone.

with the metal to form magnesium nitride. But not all of it did. As in Cavendish's case, a small bubble was left which was so inert that even hot magnesium left it cold. Ramsay measured its density and it was distinctly heavier than nitrogen. Well, was it a new element or was it merely N_3, a new and heavy form of nitrogen?

But now the spectroscope existed. The unknown gas was heated and its spectrum was observed and found to have lines that were completely new. The decision was reached at once that here was a new element. It was named "argon" from a Greek word meaning "lazy" because of its refusal to enter into any chemical combinations.

Eventually, an explanation for argon's extraordinary inertness was worked out. Each element is made up of atoms containing a characteristic number of electrons arranged in a series of shells something like the layers of an onion. To picture the situation as simply as possible, an atom is most stable when the outermost shell contains eight electrons. Chemical reactions take place in such a way that an atom either gets rid of a few electrons or takes up a few, achieving, in this way, the desired number of eight.

But what if an element has eight electrons in its outermost shell to begin with? Why, then, it is "happy" and need not react at all—and doesn't. Argon is an example. It has three shells of electrons, with the third and outermost containing eight electrons.

After argon was discovered, other examples of inert gases[2] were located; nowadays, six are known altogether: neon, with two shells of electrons; krypton, with four shells; xenon, with five shells; and radon with six shells. In each case, the outermost shell contains eight electrons. (Krypton, xenon and radon have been found, in 1962, to undergo some chemical reactions, but that is beside the present point.)

But I have mentioned only five inert gases. What of the sixth? Ah, the sixth is helium, so let's take up the helium story again.

Just before the discovery of the inert gases, in 1890, to be exact, the American chemist William Francis Hillebrand analyzed a mineral containing uranium and noticed that it gave off small quantities of an

---

[2] *This article was first written in 1960. About three years after it was written, the phrase "inert gases" went out of fashion and was replaced by "noble gases" because they proved to be not so inert after all (see Chapters 4 and 5). However, they are still the most inert substances known, so I have no hesitation in continuing to speak of them as "inert gases."*

inert gas. The gas was colorless, odorless, tasteless, insoluble, and in-combustible, so what could it be but nitrogen? He reported it as nitrogen.

When Ramsay finally came across this work some years later, he felt dissatisfied. A decision based on purely negative evidence seemed weak to him. He got hold of another uranium-containing mineral, collected the inert gas (which was there, sure enough), heated it, and studied its spectrum.

The lines were nothing like those of nitrogen. Instead, they were precisely those reported long ago by Janssen and Lockyer as having been found in sunlight. And so, in 1895, twenty-seven years after Lockyer's original assertion, the element of the sun was found on Earth. Helium did exist and it was an element. Fortunately, Janssen and Lockyer lived to see themselves vindicated. Both lived well into their eighties, Janssen dying in 1907 and Lockyer in 1920.

Helium proved interesting at once. It was the lightest of the inert gases; lighter, in fact, than any known substance but hydrogen. The helium atom had only one layer of electrons and this innermost layer can only hold two electrons. Helium has those two electrons and is therefore inert; in fact, it is the most inert of all the inert gases, and therefore of all known substances.

This extreme inertness showed up almost at once in its liquefaction point; the temperature, that is, at which it could be turned into a liquid.

When neighboring atoms (or molecules) of a substance attract each other tightly, the substance hangs together all in a piece and is solid. It can be heated to a liquid and even to a gas, the transitions coming at those temperatures where the heat energy overcomes the attractive forces between the atoms or molecules. The weaker those attractive forces, the lower the temperature required to vaporize the substance.

If the attractive force between the atoms or molecules is low enough, so little heat is required to vaporize the substance that it remains gaseous at ordinary temperatures and even, sometimes, under condi-tions of great cold.

Particularly weak attractive forces exist when atoms or molecules have the stable eight-electron arrangement in their outermost electron shells. A nitrogen molecule is composed of two nitrogen atoms which have so arranged themselves that each owns at least a share in eight electrons in its outer shell. The same is true for other simple molecules, such as those of chlorine, oxygen, carbon monoxide, hydrogen, and so on. All these are therefore gases that do not liquefy until very low temperatures are reached.

Little by little the chemists perfected their techniques for attaining low temperatures and liquefied one gas after another. Table 1 gives a measure of their progress, the liquefaction points being given in degrees Kelvin; that is, the number of Celsius degrees above absolute zero.

Table 1

GAS	YEAR FIRST LIQUEFIED	DENSITY (GRAMS PER LITER)	LIQUEFACTION POINT (° K)
Chlorine	1805	3.214	239
Hydrogen bromide	1823	3.50	206
Ethylene	1845	1.245	169
Oxygen	1877	1.429	90
Carbon monoxide	1877	1.250	83
Nitrogen	1877	1.250	77
Hydrogen	1900	0.090	20

Now, throughout the 1870s and 1880s when low-temperature work was becoming really intense, it seemed quite plain that hydrogen was going to be the hardest nut of all to crack. In general, the liquefaction point went down with density, and hydrogen was by far the least dense of all known gases and should therefore have the lowest liquefaction point. Consequently, when hydrogen was conquered, the last frontier in this direction would have fallen.

And then, just a few years before hydrogen was conquered, it lost its significance, for the inert gases had been discovered. The electronically-satisfied atoms of the inert gases had so little attraction for each other that their liquefaction points were markedly lower than other gases of similar density. You can see this in Table 2 which includes all the inert gases but helium.

As you see, radon, xenon, and krypton, all denser than chlorine, have lower liquefaction points than that gas. Argon, denser than ethylene, has a markedly lower liquefaction point than that gas; and neon, ten times as dense as hydrogen, has almost as low a liquefaction point as that lightest of all gases.

Table 2

INERT GAS	DENSITY (GRAMS PER LITER)	LIQUEFACTION POINT (° K)
Radon	9.73	211
Xenon	5.85	167
Krypton	3.71	120
Argon	1.78	87
Neon	0.90	27

The remaining inert gas, helium, which is only twice as dense as hydrogen, should, by all logic, be much more difficult to liquefy. And so it proved at once. At the temperature of liquid hydrogen, helium remained obstinately gaseous. Even when temperatures were dropped to the point where hydrogen solidified (13° K) helium remained gaseous.

It was not until 1908 that helium was liquefied. The Dutch physicist Heike Kammerlingh Onnes turned the trick; the liquefaction of helium was found to take place at 4.2° K. By allowing liquid helium to evaporate under insulated conditions, Onnes chilled it further to 1° K.

Even at 1° K, there was no sign of solid helium, however. As a matter of fact, it is now established that helium never solidifies at ordinary pressures, not even at absolute zero, where all other known substances are solid. Helium (strange element) remains liquid. There is a reasonable explanation for this. Although it is usually stated that at absolute zero all atomic and molecular motions cease, quantum mechanics shows that there is a very small residual motion that never ceases. This bit of energy suffices to keep helium liquid. To be sure, at a temperature of 1° K and a pressure of about 25 atmospheres, solid helium can be formed.

Liquid helium has something more curious to demonstrate than mere frigidity. When it is cooled below 2.2° K there is a sudden change in its properties. For one thing, helium suddenly begins to conduct heat just about perfectly. In any ordinary liquid, within a few degrees of the boiling point, there are always localized hot spots where heat happens to accumulate faster than it can be conducted away. There bubbles of vapor appear, so that there is the familiar agitation one associates with boiling.

Helium above 2.2° K ("helium I") also behaves like this. Helium below 2.2° K ("helium II"), however, vaporizes in absolute stillness, layers of atoms peeling off the top. Heat conduction is so nearly perfect that no part of the liquid can be significantly warmer than any other part and no bubbling takes place anywhere.

Furthermore, helium II has practically no viscosity. It will flow more easily than a gas and make its way through apertures that would stop a gas. It will form a layer over glass, creeping up the inner wall of a beaker and down the outer at a rate that makes it look as though it were pouring out of a hole in the beaker bottom. This phenomenon is called "superfluidity."

Odd properties are to be found in other elements at liquid helium temperatures. In 1911, Onnes was testing the electrical resistance of mercury at the temperature of liquid helium. Resistance drops with temperature, and Onnes expected resistance to reach unprecedentedly low values; but he didn't expect it to disappear altogether. Yet it did. At a temperature of 4.12° K, the electrical resistance of mercury completely vanished. This is the phenomenon of "superconductivity."

Metals other than mercury can also be made superconductive. In fact, there are a few substances that can be superconductive at nearly liquid hydrogen temperatures. Some niobium alloys become superconductive at temperatures as high as 18° K.

Superconductivity also involves an odd property with respect to a magnetic field. There are some substances that are "diamagnetic"; that is, which seem to repel magnetic lines of force. Fewer lines of force will pass through such substances than through an equivalent volume of vacuum. Well, any substance that is superconductive is completely diamagnetic; no lines of force enter it at all.

If the magnetic field is made strong enough, however, some lines of force eventually manage to penetrate the diamagnetic substance, and when that disruption of perfection takes place, all other perfections, including superconductivity, vanish. (It is odd to speak of perfection in nature. Usually perfections are the dreams of the theorist; the perfect gas, the perfect vacuum, and so on. It is only at liquid helium temperatures that true perfection seems to enter the world of reality; hence the title of this chapter.)

The phenomenon of superconductivity has allowed the invention of a tiny device that can act as a switch. In simplest form, it consists of a small wire of tantalum wrapped about a wire of niobium. If the wires are dipped in liquid helium so that the niobium wire is supercon-ductive, a tiny current passed through it will remain indefinitely, until another current is sent through the tantalum wire. The magnetic field set up in the second case, disrupts the superconductivity and stops the current in the niobium.

Properly manipulated, such a "cryotron" can be used to replace vacuum tubes or transistors. Tiny devices consisting of grouped wires, astutely arranged, can replace large numbers of bulky tubes or moder-ately bulky transistors, so that a giant computing machine of the future may well be desk-size or less if it is entirely "cryotronized."

The only catch is that for such a cryotronized computer to work, it must be dipped wholly into liquid helium.[3] The liquid helium will be vaporizing continually so that each computer will, under these condi-tions, act as an eternally continuing drain on Earth's helium supply.

Which brings up the question, of course, whether we have enough helium on Earth to support a society in which helium-dipped computers are common.

The main, and, in fact, only commercial source of the inert gases other than helium is the atmosphere, the composition of which with respect to them is given in Table 3.

Table 3

INERT GAS	PARTS PER MILLION BY WEIGHT
Argon	12,800
Neon	12.5
Krypton	2.9
Helium	0.72
Xenon	0.36
Radon	(trace)

This means that the total atmospheric content of helium is 4,500,-000,000 tons, which seems a nice tidy sum until you remember how ex-travagantly that weight of gas is diluted with oxygen and nitrogen. Helium can be obtained from liquid air, but only at tremendous ex-pense.

[3] *As a matter of fact, since this article was written, alloys have been prepared which are superconductive at temperatures in the liquid hydrogen range (over 20° K) and some optimists hope for superconductivity even at substantially higher temperatures. This may well lessen (but not entirely remove) the critical nature of the helium supply.*

(I would like to interrupt myself here to say that atmospheric helium consists almost entirely of the single isotope, helium-4. However, traces of the stable isotope, helium-3, are formed by the breakdown of radioactive hydrogen-3, which is, in turn, formed by the cosmic-ray bombardment of the atmosphere. Pure helium-3 has been studied and found to have a liquefaction point of only 3.2° K, a full degree lower than that of ordinary helium. However, helium-3 does not form the equivalent of the superfluid helium II. Only one atom of atmospheric helium per million is helium-3, so that the entire atmospheric supply amounts to only about 45,000 tons. Helium-3 is probably the rarest of all the stable isotopes here on Earth.)

But helium, at least, is found in the soil as well as in the atmosphere. Uranium and thorium give off alpha particles, which are the nuclei of helium atoms. For billions of years, therefore, helium has been slowly collecting in the Earth's crust (and remember that helium was first discovered on Earth in uranium ore and not in the atmosphere). The Earth's crust is estimated to contain helium to the extent of about 0.003 parts per million by weight. This means that the supply of helium in the crust is about twenty million times the supply in the atmosphere, but the dilution in the crust is nevertheless even greater than in the atmosphere.

However, helium is a gas. It collects in crevices and crannies and can come boiling up out of the Earth under the right conditions. In the United States, particularly, wells of natural gas often carry helium to the extent of 1 per cent, sometimes to an extent of up to 8 or even 10 per cent.

But natural gas is a highly temporary resource which we are consuming rapidly. When the gas wells peter out, so will the helium, with the only remaining supply to be found in great dilution in the atmosphere or in even greater dilution in the soil.

It is possible to imagine, then, a computerized society of the future down to its last few million cubic feet of easily-obtainable helium. What next? Scrabble for the traces in air and soil? Make do with liquid hydrogen? Abandon cryotronized computers and try to return to the giant inefficient machines of the past? Allow the culture, completely dependent on computers, to collapse?

I've been thinking about that and here is the result of my thinking.

Such a threatened society ought to have developed space travel—why not?—so that they need not seek helium only here on Earth.

Of course, the biggest source of helium in the solar system is the sun, but I see no way of snaking helium out of the sun in the foreseeable future.

The next biggest source of helium is Jupiter, which has an atmosphere that is probably thousands of miles deep, terrifically dense, and which is perhaps one-third helium by volume. Some recent theories suggest that the atmosphere is almost all helium.[4] Milking Jupiter for helium doesn't sound easy, either, but it is conceivable.

Suppose mankind could establish a base on Jupiter V, Jupiter's innermost satellite. They would then be circling a mere 70,000 miles above the visible surface of Jupiter (which is actually the upper reaches of its atmosphere). Considerable quantities of helium-loaded gas must float even higher above Jupiter than that (and therefore closer to Jupiter V).

I can imagine a fleet of unmanned ships leaving Jupiter V in a probing orbit that will carry it down toward Jupiter's surface and back, collecting and compressing gas as it does so. Such gas will be easy to separate into its components; and the helium can be liquefied far more easily out on Jupiter V than here on Earth, because the temperature is lower to begin with out there.

Uncounted tons of helium may prove easy to collect, liquefy, and store. The next logical step would be to refrain from shipping that precious stuff anywhere else, even to Earth. Why expend the energy and why undergo the tremendous losses that would be unavoidable in transit?

Instead, why not build the computers right there on Jupiter V?

And that is the vision I have, the one I mentioned at the start of the chapter. It is the vision of Jupiter V—of all places—as the nerve center of the solar system. I see this small world, a hundred miles in diameter, extracting its needed helium from the bloated world it circles and slowly being converted into one large mass of interlocking computers, swimming in the most unusual liquid that ever existed.

However, I don't think I'll have the luck of Janssen and Lockyer. Call me a pessimist, if you wish, but somehow I don't think I'll live to see this.

---

[4] *No, it isn't. Pioneer 10, streaking past Jupiter in 1973, actually detected helium in Jupiter's atmosphere but probably in minor quantities.*

4

WELCOME, STRANGER!

There are fashions in science as in everything else. Conduct an experiment that brings about an unusual success and before you can say, "There are a dozen imitations!" there are a dozen imitations!

Consider the element xenon (pronounced zee'non), discovered in 1898 by William Ramsay and Morris William Travers. Like other elements of the same type it was isolated from liquid air (see Chapter 3). The existence of these elements in air had remained unsuspected through over a century of ardent chemical analysis of the air, so when they finally dawned upon the chemical consciousness they were greeted as strange and unexpected newcomers. Indeed, the name, xenon, is the neutral form of the Greek word for "strange," so that xenon is "the strange one" in all literalness.

Xenon belongs to a group of elements commonly known as the "inert gases" (because they are chemically inert) or the "rare gases" (because they are rare), or "noble gases" because the standoffishness that results from chemical inertness seems to indicate a haughty sense of self-importance.

Xenon is the rarest of the stable inert gases and, as a matter of fact, is the rarest of all the stable elements on Earth. Xenon occurs only in the atmosphere, and there it makes up about 0.36 parts per million by weight. Since the atmosphere weighs about 5,500,000,000,000,000 (five and a half quadrillion) tons, this means that the planetary supply of xenon comes to just about 2,000,000,000 (two billion) tons. This seems ample, taken in full, but picking xenon atoms out of the overpoweringly more common constituents of the atmosphere is an arduous task and so xenon isn't a common substance and never will be.

What with one thing and another, then, xenon was not a popular substance in the chemical laboratories. Its chemical, physical, and nuclear properties were worked out, but beyond that there seemed little worth doing with it. It remained the little strange one and received cold shoulders and frosty smiles.

Then, in 1962, an unusual experiment involving xenon was announced, whereupon from all over the world broad smiles broke out across chemical countenances, and little xenon was led into the test tube with friendly solicitude. "Welcome, stranger!" was the cry everywhere, and now you can't open a chemical journal anywhere without finding several papers on xenon.

What happened?

If you expect a quick answer, you little know me. Let me take my customary route around Robin Hood's barn and begin by stating, first of all, that xenon is a gas.

Being a gas is a matter of accident. No substance is a gas intrinsically, but only insofar as temperature dictates. On Venus, water and ammonia are both gases. On Earth, ammonia is a gas, but water is not. On Titan, neither ammonia nor water are gases.

So I'll have to set up an arbitrary criterion to suit my present purpose. Let's say that any substance that remains a gas at $-100°$ C ($-148°$ F) is a Gas with a capital letter, and concentrate on those. This is a temperature that is never reached on Earth, even in an Antarctic winter of extraordinary severity, so that no Gas is ever anything but gaseous on Earth (except occasionally in chemical laboratories).

Now why is a Gas a Gas?

I can start by saying that every substance is made up of atoms, or of closely knit groups of atoms, said groups being called molecules.

There are attractive forces between atoms or molecules which make them "sticky" and tend to hold them together. Heat, however, lends these atoms or molecules a certain kinetic energy (energy of motion) which tends to drive them apart, since each atom or molecule has its own idea of where it wants to go.[1]

The attractive forces among a given set of atoms or molecules are relatively constant, but the kinetic energy varies with the temperature. Therefore, if the temperature is raised high enough, any group of atoms or molecules will fly apart and the material becomes a gas. At temperatures over 6,000° C all known substances are gases.

[1] No, I am not implying that atoms know what they are doing and have con-sciousness. This is just my teleological way of talking. Teleology is forbidden in scientific articles, but it so happens I enjoy sin.

WILLIAM RAMSAY

Ramsay, born in Glasgow, Scotland, on October 2, 1852, was an all-round man. As a youngster he was interested in music and lan-guages and then developed further interests in mathematics and sci-ence. He was an accomplished athlete as well, and to whatever he turned mind and hand, he did well. He was even a first-rate glass blower and made most of the apparatus he later used in the chemi-cal researches that brought him fame.

He studied chemistry in Germany and obtained his Ph.D. in 1873 at the University of Tübingen. In 1892 he grew intrigued by the puzzle of nitrogen—the atomic weight of nitrogen obtained from air was distinctly greater than the atomic weight of nitrogen ob-tained from chemicals.

Ramsay remembered that Cavendish, a century earlier, in a long-neglected experiment had found a small quantity of something in air that seemed not quite nitrogen. Ramsay repeated the experiment in more sophisticated fashion and obtained a bubble of left-over gas that could not be either oxygen or nitrogen. He heated it and analyzed its light by spectroscope (something Cavendish could not have done) and demonstrated it to be a new gas, denser than nitrogen and making up about 1 per cent of the atmosphere. It was com-pletely inert and was named "argon" from the Greek word for "inert."

Now that Mendeleev had worked out the periodic table, it was easy for Ramsay to see that he had discovered not one gas, but the

first of a family of gases. In 1895 he studied samples of a gas from a uranium mineral and found it was the "helium" that over thirty years before had been detected in the sun by way of its spectroscopic lines.

He then studied the argon obtained from air and, after liquefying it, allowed it to boil slowly. The first bit to boil contained still another gas called "neon" (new); the last bits contained "krypton" (hidden) and "xenon" (stranger).

Ramsay received the 1904 Nobel Prize in chemistry for his discovery of these inert gases. He died in High Wycombe, Buckinghamshire, on July 23, 1916.

Of course, there are only a few exceptional substances with interatomic or intermolecular forces so strong that it takes 6,000° C to overcome them. Some substances, on the other hand, have such weak intermolecular attractive forces that the warmth of a summer day supplies enough kinetic energy to convert them to gas (the common anethetic, ether, is an example).

Still others have intermolecular attractive forces so much weaker still that there is enough heat at a temperature of −100° C to keep them gases, and it is these that are the Gases I am talking about.

The intermolecular or interatomic forces arise out of the distribution of electrons within the atoms or molecules. The electrons are distributed among various "electron shells," according to a system we can accept without detailed explanation. For instance, the aluminum atom contains 13 electrons, which are distributed as follows: 2 in the innermost shell, 8 in the next shell, and 3 in the next shell. We can therefore signify the electron distribution in the aluminum atom as 2,8,3.

The most stable and symmetrical distribution of the electrons among the electron shells is that distribution in which the outermost shell holds either all the electrons it can hold, or 8 electrons—whichever is less. The innermost electron shell can hold only 2, the next can hold 8, and each of the rest can hold more than 8. Except for the situation where only the innermost shell contains electrons, then, the stable situation consists of 8 electrons in the outermost shell.

There are exactly six elements known (see Table 4) in which this situation of maximum stability exists:

Table 4

ELEMENT	SYMBOL	ELECTRON DISTRIBUTION	ELECTRON TOTAL
Helium	He	2	2
Neon	Ne	2,8	10
Argon	Ar	2,8,8	18
Krypton	Kr	2,8,18,8	36
Xenon	Xe	2,8,18,18,8	54
Radon	Rn	2,8,18,32,18,8	86

Other atoms without this fortunate electronic distribution are forced to attempt to achieve it by grabbing additional electrons, or getting rid of some they already possess, or sharing electrons. In so doing, they undergo chemical reactions. The atoms of the six elements listed above, however, need do nothing of this sort and are sufficient unto themselves. They have no need to shift electrons in any way and that means they take part in no chemical reactions and are inert. (At least, this is what I would have said prior to 1962.)

The atoms of the inert gas family listed above are so self-sufficient, in fact, that the atoms even ignore one another. There is little interatomic attraction, so that all are gases at room temperature and all but radon are Gases.

To be sure, there is *some* interatomic attraction (for no atoms or molecules exist among which there is no attraction at all). If one lowers temperature sufficiently, a point is reached where the attractive forces become dominant over the disruptive effect of kinetic energy, and every single one of the inert gases will, eventually, become an inert liquid.

What about other elements? As I said, these have atoms with electron distributions of less than maximum stability and each has a tendency to alter that distribution in the direction of stability. For instance, the sodium atom (Na) has a distribution of 2,8,1. If it could get rid of the outermost electron, what would be left would have the stable 2,8 configuration of neon. Again, the chlorine atom (Cl) has a distribution of 2,8,7. If it could gain an electron, it would have the 2,8,8 distribution of argon.

Consequently, if a sodium atom encounters a chlorine atom, the transfer of an electron from the sodium atom to the chlorine atom satisfies both. However, the loss of a negatively charged electron leaves the sodium atom with a deficiency of negative charge or, which is the same thing, an excess of positive charge. It becomes a positively charged sodium ion (Na^+). The chlorine atom, on the other hand, gaining an electron, gains an excess of negative charge and becomes a negatively charged chloride ion[2] (Cl^-).

Opposite charges attract, so the sodium ion attracts all the chloride ions within reach and vice versa. These strong attractions cannot be overcome by the kinetic energy induced at ordinary temperatures, and so the ions hold together firmly enough for "sodium chloride"

---

[2] *The charged chlorine atom is called "chloride ion" and not "chlorine ion" as a convention of chemical nomenclature we might just as well accept with a weary sigh. Anyway, the "d" is not a typographical error.*

(common salt) to be a solid. It does not become a gas, in fact, until a temperature of 1,413° C is reached.

Next, consider the carbon atom (C). Its electron distribution is 2,4. If it lost 4 electrons, it would gain the 2 helium configuration; if it gained 4 electrons, it would gain the 2,8 neon configuration. Losing or gaining that many electrons is not easy, so the carbon atom shares electrons instead. It can, for instance, contribute one of its electrons to a "shared pool" of two electrons, a pool to which a neighboring carbon atom also contributes an electron. With its second electron it can form another shared pool with a second neighbor, and with its third and fourth, two more pools with two more neighbors. Each neighbor can set up additional pools with other neighbors. In this way, each carbon atom is surrounded by four other carbon atoms.

These shared electrons fit into the outermost electron shells of each carbon atom that contributes. Each carbon atom has 4 electrons of its own in that outermost shell and 4 electrons contributed (one apiece) by four neighbors. Now, each carbon atom has the 2,8 configuration of neon, but only at the price of remaining close to its neighbors. The result is a strong interatomic attraction, even though electrical charge is not involved. Carbon is a solid and is not a gas until a temperature of 4,200° C is reached.

The atoms of metallic elements also stick together strongly, for similar reasons, so that tungsten, for instance, is not a gas until a temperature of 5,900° C is reached.

We cannot, then, expect to have a Gas when atoms achieve stable electron distribution by transferring electrons in such a manner as to gain an electric charge; or by sharing electrons in so complicated a fashion that vast numbers of atoms stick together in one piece.

What we need is something intermediate. We need a situation where atoms achieve stability by sharing electrons (so that no electric charge arises) but where the total number of atoms involved in the sharing is very small so that only small molecules result. Within the molecules, attractive forces may be large, and the molecules may not be shaken apart without extreme temperature. The attractive forces between one molecule and its neighbor, however, may be small—and that will do.

Let's consider the hydrogen atom, for instance. It has but a single electron. Two hydrogen atoms can each contribute its single electron to form a shared pool. As long as they stay together, each can count both electrons in its outermost shell and each will have the stable helium configuration. Furthermore, neither hydrogen atom will have any electrons left to form pools with other neighbors, hence the molecule will end there. Hydrogen gas will consist of two-atom molecules (H_2).

The attractive force between the atoms in the molecule is large, and it takes temperatures of more than $2,000°$ C to shake even a small fraction of the hydrogen molecules into single atoms. There will, however, be only weak attractions among separate hydrogen molecules, each of which, under the new arrangement, will have reached a satisfactory pitch of self-sufficiency. Hydrogen, therefore, will be a Gas—one not made up of separate atoms as is the case with the inert gases, but of two-atom molecules.

Something similar will be true in the case of fluorine (electronic distribution 2,7), oxygen (2,6) and nitrogen (2,5). The fluorine atom can contribute an electron and form a shared pool of two electrons with a neighboring fluorine atom which also contributes an electron. Two oxygen atoms can contribute two electrons apiece to form a shared pool of four electrons, and two nitrogen atoms can contribute three electrons each and form a shared pool of six electrons.

In each case, the atoms will achieve the 2,8 distribution of neon at the cost of forming paired molecules. As a result, enough stability is achieved so that fluorine (F_2), oxygen (O_2), and nitrogen (N_2) are all Gases.

The oxygen atom can also form a shared pool of two electrons with each of two neighbors, and those two neighbors can form another shared pool of two electrons among themselves. The result is a combination of three oxygen atoms (O_3), each with a neon configuration. This combination, O_3, is called ozone, and it is a Gas too.

Oxygen, nitrogen, and fluorine can form mixed molecules, too. For instance, a nitrogen and an oxygen atom can combine to achieve the necessary stability for each. Nitrogen may also form shared pools of two electrons with each of three fluorine atoms, while oxygen may do so with each of two. The resulting compounds: nitrogen oxide (NO), nitrogen trifluoride (NF_3), and oxygen difluoride (OF_2) are all Gases.

Atoms which, by themselves, will not form Gases may do so if combined with either hydrogen, oxygen, nitrogen, or fluorine. For instance, two chlorine atoms (2,8,7, remember) will form a shared pool of two electrons so that both achieve the 2,8,8 argon configuration. Chlorine (Cl_2) is therefore a gas at room temperature—with intermolecular attractions, however, large enough to keep it from being a Gas. Yet if a chlorine atom forms a shared pool of two electrons with a fluorine atom, the result, chlorine fluoride (ClF), *is* a Gas.

The boron atom (2,3) can form a shared pool of two electrons with each of three fluorine atoms, and the carbon atom a shared pool of two electrons with each of four fluorine atoms. The resulting compounds, boron trifluoride (BF_3) and carbon tetrafluoride (CF_4), are Gases.

A carbon atom can form a shared pool of two electrons with each of four hydrogen atoms, or a shared pool of four electrons with an oxygen atom, and the resulting compounds, methane (CH_4) and carbon monoxide (CO), are gases. A two-carbon combination may set up a shared pool of two electrons with each of four hydrogen atoms (and a shared pool of four electrons with one another); a silicon atom may set up a shared pool of two electrons with each of four hydrogen atoms. The compounds, ethylene (C_2H_4) and silane (SiH_4), are Gases.

Altogether, then, I can list twenty Gases which fall into the following categories:

(1) Five elements made up of single atoms: helium, neon, argon, krypton, and xenon.

(2) Four elements made up of two-atom molecules: hydrogen, nitrogen, oxygen, and fluorine.

(3) One element form made up of three-atom molecules: ozone (of oxygen).

(4) Ten compounds, with molecules built up of two different elements, at least one of which falls into category (2).

The twenty Gases are listed in order of increasing boiling point (B.P.) in Table 5, and that boiling point is given in both the Celsius scale (° C) and the Absolute scale (° K).

The five inert gases on the list are scattered among the fifteen other Gases. To be sure, two of the three lowest-boiling Gases are helium and neon, but argon is seventh, krypton is tenth, and xenon is seventeenth. It would not be surprising if all the Gases, then, were as inert as the inert gases.

Table 5

THE TWENTY GASES

SUBSTANCE	FORMULA	B.P. (°C)	B.P. (°K)
Helium	He	−268.9	4.2
Hydrogen	H_2	−252.8	20.3
Neon	Ne	−245.9	27.2
Nitrogen	N_2	−195.8	77.3
Carbon monoxide	CO	−192	81
Fluorine	F_2	−188	85
Argon	Ar	−185.7	87.4
Oxygen	O_2	−183.0	90.1
Methane	CH_4	−161.5	111.6
Krypton	Kr	−152.9	120.2
Nitrogen oxide	NO	−151.8	121.3
Oxygen difluoride	OF_2	−144.8	128.3
Carbon tetrafluoride	CF_4	−128	145
Nitrogen trifluoride	NF_3	−120	153
Ozone	O_3	−111.9	161.2
Silane	SiH_4	−111.8	161.3
Xenon	Xe	−107.1	166.0
Ethylene	C_2H_4	−103.9	169.2
Boron trifluoride	BF_3	−101	172
Chlorine fluoride	ClF	−100.8	172.3

Perhaps they might be at that, if the smug, self-sufficient molecules that made them up were permanent, unbreakable affairs, but they are not. All the molecules can be broken down under certain conditions, and the free atoms (those of fluorine and oxygen particularly) are active indeed.

This does not show up in the Gases themselves. Suppose a fluorine molecule breaks up into two fluorine atoms, and these find themselves surrounded only by fluorine molecules? The only possible result is the re-formation of a fluorine molecule, and nothing much has happened. If, however, there are molecules other than fluorine present, a new molecular combination of greater stability than F_2 is possible (indeed, almost certain in the case of fluorine), and a chemical reaction results.

The fluorine molecule does have a tendency to break apart (to a very small extent) even at ordinary temperatures, and this is enough. The free fluorine atom will attack virtually anything non-fluorine in sight, and the heat of reaction will raise the temperature, which will bring about a more extensive split in fluorine molecules, and so on. The result is that molecular fluorine is the most chemically active of all the Gases (with chlorine fluoride almost on a par with it and ozone making a pretty good third).

The oxygen molecule is torn apart with greater difficulty and therefore remains intact (and inert) under conditions where fluorine will not. You may think that oxygen is an active element, but for the most part this is only true under elevated temperatures, where more energy is available to tear it apart. After all, we live in a sea of free oxygen without damage. Inanimate substances such as paper, wood, coal, and gasoline, all considered flammable, can be bathed by oxygen for indefinite periods without perceptible chemical reaction—until heated.

Of course, once heated, oxygen does become active and combines easily with other Gases such as hydrogen, carbon monoxide, and methane which, by that token, can't be considered particularly inert either.

The nitrogen molecule is torn apart with still more difficulty and, before the discovery of the inert gases, nitrogen was *the* inert gas *par excellence*. It and carbon tetrafluoride are the only Gases on the list, other than the inert gases themselves, that are respectably inert, but even they can be torn apart.

Life depends on the fact that certain bacteria can split the nitrogen molecule; and important industrial processes arise out of the fact that man has learned to do the same thing on a large scale. Once the nitrogen molecule is torn apart, the individual nitrogen atom is quite active, bounces around in all sorts of reactions and, in fact, is the fourth most common atom in living tissue and is essential to all its workings.

In the case of the inert gases, all is different. There are no molecules to pull apart. We are dealing with the self-sufficient atom itself, and there seemed little likelihood that combination with any other atom would produce a situation of greater stability. Attempts to get inert gases to form compounds, at the time they were discovered, failed, and chemists were quickly satisfied that this made sense.

To be sure, chemists continued to try, now and again, but they also continued to fail. Until 1962, then, the only successes chemists had had in tying the inert gas atoms to other atoms was in the formation of "clathrates." In a clathrate, the atoms making up a molecule form a cage-like structure and, sometimes, an extraneous atom—even an inert gas atom—is trapped within the cage as it forms. The inert gas is then tied to the substance and cannot be liberated without breaking down the molecule. However, the inert gas atom is only physically confined; it has not formed a chemical bond.

And yet, let's reason things out a bit. The boiling point of helium is 4.2° K; that of neon is 27.2° K, that of argon 87.4° K, that of krypton 120.2° K, that of xenon 166.0° K. The boiling point of radon, the sixth and last inert gas and the one with the most massive atom, is 211.3° K (−61.8° C). Radon is not even a Gas, but merely a gas.

Furthermore, as the mass of the inert gas atoms increases, the ionization potential (a quantity which measures the ease with which an electron can be removed altogether from a particular atom) decreases. The increasing boiling point and decreasing ionization potential both indicate that the inert gases become less inert as the mass of the individual atoms rises.

By this reasoning, radon would be the least inert of the inert gases and efforts to form compounds should concentrate upon it as offering the best chance. However, radon is a radioactive element with a half-life of less than four days, and is so excessively rare that it can be worked with only under extremely specialized conditions. The next best bet, then, is xenon. This is very rare, but it is available and it is, at least, stable.

Then, if xenon is to form a chemical bond, with what other atom might it be expected to react? Naturally, the most logical bet would be to choose the most reactive substance of all—fluorine or some fluorine-containing compound. If xenon wouldn't react with that, it wouldn't react with anything.

(This may sound as though I am being terribly wise after the event, and I am. However, there are some who were legitimately wise. I am told that Linus Pauling reasoned thus in 1932, well before the event, and that a gentleman named A. von Antropoff did so in 1924.)

In 1962, Neil Bartlett and others at the University of British Columbia were working with a very unusual compound, platinum hexafluoride (PtF_6). To their surprise, they discovered that it was a particularly active compound. Naturally, they wanted to see what it could be made to do, and one of the thoughts that arose was that here might be something that could (just possibly) finally pin down an inert gas atom.

So Bartlett mixed the vapors of PtF_6 with xenon and, to his astonishment, obtained a compound which seemed to be $XePtF_6$, xenon platinum hexafluoride. The announcement of this result left a certain area of doubt, however. Platinum hexafluoride was a sufficiently complex compound to make it just barely possible that it had formed a clathrate and trapped the xenon.

A group of chemists at Argonne National Laboratory in Chicago therefore tried the straight xenon-plus-fluorine experiment, heating one part of xenon with five parts of fluorine under pressure at $400°$ C in a nickel container. They obtained xenon tetrafluoride (XeF_4), a straightforward compound of an inert gas, with no possibility of a clathrate. (To be sure, this experiment could have been tried years before, but it is no disgrace that it wasn't. Pure xenon is very hard to get and pure fluorine is very dangerous to handle, and no chemist could reasonably have been expected to undergo the expense and the risk for so slimchanced a catch as an inert gas compound until after Bartlett's experiment had increased that "slim chance" tremendously.)

And once the Argonne results were announced, all Hades broke loose. It looked as though every inorganic chemist in the world went gibbering into the inert gas field. A whole raft of xenon compounds, including not only XeF_4, but also XeF_2, XeF_6, $XeOF_2$, $XeOF_3$, $XeOF_4$, XeO_3, H_4XeO_4, and H_4XeO_6 have been reported.

Enough radon was scraped together to form radon tetrafluoride (RnF_4). Even krypton, which is more inert than xenon, has been tamed, and krypton difluoride (KrF_2) and krypton tetrafluoride (KrF_4) have been formed.

The remaining three inert gases, argon, neon, and helium (in order of increasing inertness), as yet remain untouched. They are the last of the bachelors, but the world of chemistry has the sound of wedding bells ringing in its ears, and it is a bad time for bachelors.[3]

---

[3] *This article was written in 1963. Now, eleven years later, the three bachelors are* still *bachelors.*

5

DEATH IN THE LABORATORY

I'm a great one for iconoclasm. Given half a chance, I love to say something shattering about some revered institution, and wax sarcastically cynical about Mother's Day or apple pie or baseball. Naturally, though, I draw the line at having people say nasty things about institutions I personally revere.

Like Science and Scientists, for instance. (Capital S, you'll notice.)

Scientists have their faults, of course. They can be stodgy and authoritarian and theories can get fixed in place and resist dislodging. There is, for instance, the sad case of the French chemist Auguste Laurent and the Swedish chemist Jöns Jakob Berzelius.

In 1836, Laurent advanced theories concerning the structure of organic compounds that were on the right track, while Berzelius had long maintained views in this respect that had important elements of wrongness. Unfortunately, Laurent was young and little known and Berzelius was the Great Man of chemistry in his time, so Laurent was hounded into obscurity. He was forced to work in third-class poorly heated laboratories, since no important institution would hire him in the face of

Berzelius' displeasure, and the poor working conditions aggravated his tubercular condition and brought him to a premature death. Berzelius, on the other hand, died at the peak of his fame and it was only after his death that Laurent's views began to win out.

These things happen, alas, but not as often in science (I like to think) as in any other form of human endeavor.

At any rate, if someone is going to berate Science as an organization in which Authority stifles Initiative, and in which Vain Old Men squash Eager Young Geniuses, and where the lack of the union card of the Ph.D. condemns brilliant amateurs to the outer darkness—it would be nice if some legitimate examples were used.

Occasionally someone treats the discovery of xenon fluoride (see Chapter 4) as an example of the manner in which stodgy theories actually inhibit experimentation.

I can hear them say it: "Stupid lazy chemists just got the idea into their heads that the noble gases formed no compounds so no one bothered to try to see if they *could* form compounds. After all, if everyone *knows* that something can't be done, why try to do it? And yet, if, at any time, any chemist had simply bothered to mix xenon and fluorine in a nickel container—"

It does sound very stupid of a chemist not to stumble on something that easy, doesn't it? Just mix a little xenon and fluorine in a nickel container, and astonish the world, and maybe win a Nobel Prize.

But do you know what would have happened if the average chemist in the average laboratory had tired to mix a little xenon (very rare and quite expensive, by the way) with a little fluorine? A bad case of poisoning, very likely, and, quite possibly, death.

If you think I'm exaggerating, let's consider the history of fluorine. That history does not begin with fluorine itself—a pale yellow-green assassin never seen by human eyes until eighty years ago—but with an odd mineral used by German miners about five hundred years ago.

The substance is mentioned by the first great mineralogist of modern times George Agricola. In 1529, he described its use by German miners. The mineral melted easily (for a mineral) and when added to ore being smelted, the entire mixture melted more easily, thus bringing about a valuable saving of fuel and time.

Something which is liquid flows, and the Latin word for "to flow" is *fluere*, from which we get "fluid" (for any substance that is a liquid or gas and flows) and "fluent" (to describe an easy flow of words). From the same root comes the word Agricola used for the mineral that liquefied and flowed so easily. That word was *fluores*.

In later years, it came to be called "fluorspar," since "spar" is an old miners' term for "rock." Then, when it became customary to add the suffix "-ite" to the names of minerals, a new alternate name was "fluorite." (The name had an important descendant when it was discovered that fluorite, upon exposure to light of one wavelength, gave off light of a longer wavelength. That process came to be known as "fluorescence.")

Fluorite is still used today as a flux (or liquefier) in the making of steel. The centuries pass but a useful property remains a useful property.

In 1670, a German glass-cutter, Heinrich Schwanhard, was working with fluorite and exposing it, for some reason, to the action of strong acids. A vapor was given off and Schwanhard bent close to watch. His spectacles clouded and, presumably, he may have thought the vapor had condensed upon them.

The cloud did not disappear, however, and on closer examination, the spectacles proved to have been etched. The glass had actually been partly dissolved and its smooth surface roughened.

This was very unusual, for few chemicals attack glass, which is one of the reasons chemists use glassware for their equipment. Schwanhard saw a Good Thing is this. He learned to cover portions of glass objects with wax (which protected those portions against the vapors) and etched the rest of the glass. In this way, he formed all sorts of delicate figures in clear glass against a cloudy background. He got himself patronized by the Emperor and did very well, indeed.

But he kept his process secret and it wasn't until 1725 that chemists, generally, learned of this interesting vapor.

Through the eighteenth century, there were occasional reports on fluorite. A German chemist, Andreas Sigismund Marggraf, showed, in 1768, that fluorite did not contain sulfur. He also found that fluorite, treated with acid, produced a vapor that chewed actual holes in his glassware.

However, it was a Swedish chemist, Carl Wilhelm Scheele, who really put the glass-chewing gas on the map about 1780. He, too, acidified fluorite and etched glass. He studied the vapors more thoroughly than

CARL WILHELM SCHEELE

Scheele was born in Stralsund, Pomerania, a region which was German through most of history, but happened to be Swedish at the time of his birth on December 9, 1742. Although Scheele was of German descent, he lived and worked in Sweden and is usually considered a Swedish chemist.

He was the seventh child of eleven, and at the age of fourteen he was apprenticed to an apothecary. In those days, apothecaries were profoundly interested in minerals and usually prepared their own drugs; Scheele taught himself chemistry and became probably the greatest apothecary in history. Later in life he could easily have obtained a university position, with all its prestige, but he preferred to remain an apothecary and concentrate on research.

In the course of his work he discovered a number of acids including tartaric acid, citric acid, lactic acid, and uric acid. He prepared and any predecessor and maintained the gas to be an acid. Because of this, Scheele is commonly given the credit for having discovered this "fluoric acid" (as it was termed for about a quarter of a century).

The discovery, unfortunately, did Scheele's health no good. He isolated a large number of substances and it was his habit to smell and taste all the new chemicals he obtained, in order that this might serve as part of the routine characterization. Since in addition to the dangerous "fluoric acid," he also isolated such nasty items as hydrogen sulfide (the highly poisonous rotten-egg gas we commonly associate with school chemistry laboratories) and hydrogen cyanide (used in gas-chamber executions), the wonder is that he didn't die with the stuff in his mouth.

His survival wasn't total, though, for he died at the early age of forty-three, after some years of invalidship. There is no question in my mind but that his habit of sniffing and sipping unknown chemicals drastically shortened his life.

While most chemists are very careful about tasting, by the way, much more careful than poor Scheele ever was, this cannot be said about smelling, even today. Chemists may not deliberately go about sniffing at things, but the air in laboratories is usually loaded with gases and vapors and chemists often take a kind of perverse pleasure in tolerating

investigated three highly poisonous gases, hydrogen fluoride, hydrogen sulfide, and hydrogen cyanide, and managed to avoid killing himself (though he died at the age of forty-three, his life surely shortened by the damage he did himself in smelling and tasting new chemicals in order to characterize them). He even recorded the taste of hydrogen cyanide—a report one would swear could only be made posthumously.

He was involved in the discovery of seven different elements, yet did not manage to obtain the credit for the discovery of any one of them. The most tragic case was that of oxygen, which he prepared in 1771, three years before Priestley did. Through the criminal neglect of his publisher, Scheele's report of his discovery was not printed till 1777 and Priestley, publishing first, got the credit.

He eschewed virtually all social life in favor of science, his only passion, and when he decided to marry he found he had time for it only on his deathbed. He died in Köping, Sweden, on May 21, 1786.

this, and in reacting with a kind of superior professional amusement at the nonchemists who make alarmed faces and say "phew."

This may account for the alleged shortened life expectancy of chemists generally. I am not speaking of this shortened life expectancy as an established fact, please note, since I don't know that it is. I say "alleged." Still, there was a letter recently in a chemical journal by someone who had been following obituaries and who claimed that chemists died at a considerably younger age, on the average, than did scientists who were not chemists. This could be so.

There were also speculations some years back that a number of chemists showed mental aberrations in later years through the insidious long-term effects of mercury poisoning. This came about through the constant presence of mercury vapor in the laboratory, vapor that ascended from disregarded mercury droplets in cracks and corners. (All chemists spill mercury now and then.)

To avoid creating alarm and despondency, however, I might mention that some chemists lived long and active lives.[1] The prize specimen is the French chemist Michel Eugène Chevreul, who was born in 1786 and died in 1889 at the glorious age of one hundred and three! What's more, he was active into advanced old age, for in his nineties he was making useful studies on gerontology (the study of the effect of old age on living organisms) using himself (who else) as a subject. He attended the elaborate celebration of the centennial of his own birth and was exuberantly hailed as the "Nestor of science." Indeed, I know of no other scientist of the first class who passed the age of one hundred.[2] If a Gentle Reader knows of one, please let me have the information.

Of course, Chevreul worked with such nondangerous substances as waxes, soaps, fats, and so on, but consider then the German chemist Robert Wilhelm Bunsen. As a young man he worked with organic compounds of arsenic and poisoned himself nearly to the point of death. At the age of twenty-five, one of those compounds exploded and caused him to lose the sight of one eye. He survived, however, and went on to attain the respectable age of eighty-eight.

Yet it remains a fact that many of the chemists, in the century after

---

[1] And to those loyal Readers who may be concerned about my personal welfare, I must admit that for many years now I have entered chemistry laboratories only at rare intervals.

[2] On October 23, 1973, William David Coolidge, who first developed the tungsten wires in light bulbs, lived to see his hundredth birthday.

Scheele, who did major work in "fluoric acid" died comparatively young.

Once Scheele had established the gas produced from acidified fluorite to be an acid, a misconception at once arose as to its structure. The great French chemist Antoine Laurent Lavoisier had decided at just about that time that all acids contained oxygen and, indeed, the word "oxygen" is from the Greek phrase meaning "acid producer."

It is true that many acids contain oxygen (sulfuric acid and nitric acid are examples) but some do not. Consider, for instance, a compound called "muriatic acid," from a Latin word meaning "brine" because the acid could be obtained by treating brine with sulfuric acid.

It was supposed, following Lavoisier's dictum, that muriatic acid contained oxygen, was perhaps a compound of oxygen with an as-yet-unknown element called "murium." Scheele found that on treating muriatic acid with certain oxygen-containing compounds, a greenish gas was obtained. He assumed that muriatic acid had added on additional oxygen and named the gas "oxymuriatic acid."

The English chemist Humphry Davy, however, after careful work with muriatic acid was able to show that the acid did not contain oxygen. Rather it contained hydrogen and was probably a compound of hydrogen and an as-yet-unknown element. Furthermore, if oxygen combined with muriatic acid, the chances were that it combined with the hydrogen, pulling it away and leaving the as-yet-unknown element in isolation. The greenish gas which Scheele had called oxymuriatic acid was, Davy decided, that element, and in 1810 he renamed it "chlorine" from the Greek word for "green" because of its color.

Since muriatic acid is a compound of hydrogen and chlorine, it came to be known as "hydrogen chloride" (in gaseous form) or "hydrochloric acid" (in water solution).

Other acids were also found to be free of oxygen. Hydrogen sulfide and hydrogen cyanide are examples. (They are very weak acids, to be sure, but the oxygen-in-acid proponents could not fall back on the assumption that oxygen is required for *strong* acids, since hydrochloric acid, though not containing oxygen, is, nevertheless, a strong acid.)

Davy went on to show that fluoric acid was another example of an acid without oxygen. Furthermore, fluoric acid had certain properties that were quite reminiscent of hydrogen chloride. It occurred to a French physicist, André Marie Ampère, therefore, that fluoric acid might well be a compound of hydrogen with an element very like chlorine. He said as much to Davy, who agreed.

By 1813, Ampère and Davy were giving the new element (not yet isolated or studied) the same suffix as that possessed by chlorine in order to emphasize the similarity. The stem of the name would come from fluorite, of course, and the new element was "fluorine," a name that has been accepted ever since. Fluoric acid became "hydrogen fluoride" and fluorite became "calcium fluoride."

The problem now arose of isolating fluorine so that it might be studied. This proved to be a problem of the first magnitude. Chlorine could be isolated from hydrochloric acid by having oxygen, so to speak, snatch the hydrogen from chlorine's grip, leaving the latter isolated as the element. Oxygen was more active than chlorine, you see, and pulled more strongly at hydrogen than chlorine could.

The same procedure could not, however, be applied to hydrogen fluoride. Oxygen could not, under any conditions, snatch hydrogen from the grip of fluorine. (It was found, many years later, that elementary fluorine could, instead, snatch hydrogen from oxygen. Fluorine, in reacting with water—a compound of hydrogen and oxygen—snatches at the hydrogen with such force that the oxygen is liberated in the unusually energetic form of ozone.)

HUMPHRY DAVY

Davy was born poor, on December 17, 1778, in Penzance, Cornwall, his father having left only a large debt as legacy (though Davy and his mother eventually paid it off in full). Davy did not enjoy school and was soon apprenticed to an apothecary, where he began a program of self-education.

His interests were at first rather wide-ranging. He was an enthusiastic fisherman and wrote a book on the subject. He was interested in philosophy and was considered to have displayed considerable talent as a poet. He gave lectures later in life with the poise and charm of a born showman, and to top it off he was very handsome. (Some historians of science consider him to have been the handsomest of the great scientists.)

He experimented with gases early in life and studied their effects on being inhaled. He breathed four quarts of hydrogen, for instance, nearly to the point of his own suffocation and tried to breathe pure carbon dioxide. He discovered nitrous oxide in 1800 and found that,

when inhaled, it gave him a giddy, intoxicated feeling. Others tried it and found that under its influence, inhibitions were lowered so that the breathers laughed and cried easily. For a while, nitrous oxide ("laughing gas") parties were organized. It was the mind-expander of its time.

Davy achieved fame when he took on the job of lecturer for the Royal Institution. The Napoleonic Wars were preventing the gentry from traveling to the Continent and the ladies, particularly, flocked to hear the handsome young scientist.

A greater fame, though, came when he devised methods for passing electricity through molten compounds and isolating, for the first time, a whole series of metals such as sodium, potassium, calcium, barium, magnesium, and strontium.

Yet there is no doubt that the greatest of all his discoveries was Michael Faraday, the pupil who proved to be greater than the teacher, and of whom Davy was fiendishly jealous. Davy died in Geneva, Switzerland, May 29, 1829.

The conclusion was inescapable that fluorine was more active than chlorine and oxygen. In fact, there seemed reason to suspect that fluorine might be the most active element in existence (a deduction that later chemists amply confirmed) and that no simple chemical reaction could liberate fluorine from compounds such as hydrogen fluoride or calcium fluoride, since no other element could force hydrogen or calcium out of the strong grip of fluorine.

But then, who says it is necessary to restrict one's self to chemical reactions. In 1800, the electric battery was invented and within weeks, it had been found that an electric current passing through a compound could split it apart ("electrolysis") where ordinary chemical reactions might be able to perform that task only under extreme conditions. Water, for instance, was broken up to hydrogen and oxygen. Hydrogen (and various metals) can be made to appear at the negative electrode, while oxygen (and other nonmetals) can be made to appear at the positive electrode.

Davy applied this technique to various compounds which chemists were sure contained still-unknown metals that were so active that ordinary chemical techniques did not suffice to break them loose. In 1807 and 1808, making use of the most powerful electric battery that had yet been constructed, he quickly isolated six extremely active metals: potassium, sodium, calcium, magnesium, strontium, and barium. All appeared at the negative electrode, of course.

There was no reason, it seemed to Davy, that the same technique might not work with calcium fluoride. Here the calcium would appear at the negative electrode and fluorine at the positive. He tried it and got nowhere. Oh, he might have isolated fluorine at the positive electrode, but as soon as it was formed, it attacked whatever was in sight: water, glass, even silver or platinum, which Davy had used as his container. In no time at all, Davy had fluorine compounds on his hands, but no fluorine.

It was a losing proposition in another way, too, for Davy managed to be severely poisoned during his work on fluorine compounds, through breathing small quantities of hydrogen fluoride. It didn't kill him, but it undoubtedly contributed to the fact that he died at the age of fifty after some years of invalidship.

Others were less lucky than Davy, even. In the 1830s, two English brothers, Thomas and George Knox, decided not to take it for granted that fluorine could not be liberated by chemical means (scientists are not as stodgy as their critics like to pretend). They tried to coax chlorine into reacting with mercury fluoride, accepting the mercury and liberating the fluorine. They failed, and both underwent long and agonizing sieges of hydrogen fluoride poisoning.

A Belgian chemist, P. Louyet, who had followed the attempts of the Knox brothers closely, tried to repeat their work and failed even more spectacularly. He was entirely killed by hydrogen fluoride.

One of Louyet's assistants was the French chemist Edmond Frémy. He had watched some of Louyet's experiments and decided that trying to isolate fluorine by chemical reactions got one nothing but a ticket to the morgue. He returned to Davy's electrolytic method and worked with the most gingerly caution. His reward was that he lived to be eighty.

In 1885, he repeated Davy's attempt to electrolyze calcium fluoride with the same results—any fluorine that developed tackled everything in reach and was gone at once.

He next decided to work with hydrogen fluoride itself. Hydrogen fluoride is a liquid at slightly less than room temperatures and can be more easily dealt with. It needn't be kept red-hot during the electrolysis as calcium fluoride has to be.

Unfortunately, hydrogen fluoride in Frémy's day was always obtained in water solution. To try to electrolyze a water solution of hydrogen flouride meant that two different elements could come off at the positive electrode, oxygen or fluorine. Since oxygen was less active and more easily pulled away from hydrogen, only oxygen appeared at the electrode if there was even a small quantity of water present in the hydrogen fluoride.

Frémy therefore worked out methods for producing completely water-free hydrogen fluoride: "anhydrous hydrogen fluoride." He was the first to do so. Unfortunately, he found himself stymied. Anhydrous hydrogen fluoride would not pass an electric current. If he added some water, an electric current would pass—but only oxygen would be produced.

In the end, he, too, gave up and as the 1880s dawned, fluorine was still victor. It had defeated the best efforts of many first-class chemists for three-quarters of a century; had invalided some and killed others outright.

FERDINAND FRÉDÉRIC HENRI MOISSAN

Moissan was born in Paris on September 28, 1852, and his early schooling was hampered by poverty. At the age of eighteen he was apprenticed to an apothecary. His interest in chemistry was great enough, however, for him to break away two years later and obtain by hard labor the education he believed he needed. In 1882 he married, for love, a young lady who happened to have a well-to-do father who was willing to support the young man while he studied for his Ph.D., which he obtained in 1885.

Moissan studied under Edmond Frémy who was interested in isolating fluorine, and Moissan took on the task. In his effort, he made use of platinum, one of the few materials that were reasonably immune to the onslaughts of fluorine. On June 26, 1886, he passed an electric current through a solution of potassium fluoride in hydrofluoric acid in all-platinum equipment which was chilled to −50° C. A pale yellow gas was produced that was the long-sought fluorine.

Frémy had a student, the French chemist Ferdinand Frédéric Henri Moissan, who took up the battle, and proceeded to attack the fluorine problem with bulldog tenacity.

He went back to chemical methods once again. He decided he must begin with a fluorine compound that was relatively unstable. The more stable a compound after all, the more tightly fluorine is holding the other atoms and the more difficult it is to pry that fluorine loose.

In 1884, Moissan came to the conclusion that phosphorus fluoride was comparatively unstable (for a fluoride). This seemed particularly hopeful since phosphorus happened to be unusually avid in its tendency to combine with oxygen. Perhaps in this case, oxygen could pull atoms away from fluorine. Moissan tried and succeeded only partially. The oxygen grabbed at the phosphorus all right but the fluorine did not let go and Moissan ended with a compound in which phosphorus was combined with both oxygen and fluorine.

Moissan tried another tack. Platinum is an extremely inert metal; even fluorine attacks it only with difficulty. Hot platinum, however, does seem to have the ability to combine easily with phosphorus. If he passed phosphorus fluoride over hot platinum, would the platinum perhaps combine with phosphorus rather than with fluorine, and set the fluorine free?

No such luck. Both phosphorus and fluorine combined with the

In 1906 Moissan won the Nobel Prize in chemistry for this feat, winning the award, according to report, by only one vote over Mendeleev, the inventor of the periodic table.

Moissan was also interested in trying to convert carbon from its usual form (graphite) into the much more uncommon form of diamond. For this he made use of high temperatures and pressures and in 1893 it seemed to him he had succeeded. Several tiny diamonds were reported by him, and a sliver of colorless diamond, over half a millimeter in length, was exhibited. We now know that Moissan couldn't possibly have attained the temperatures and pressures needed for the task, and some think he was hoaxed by some assistants as a practical joke which went too far and proved too embarrassing to own up to.

Moissan died in Paris on February 20, 1907, at the age of fifty-four, possibly because of the damage to his health produced by the fluorine compounds he worked with.

platinum and in a matter of minutes a lot of expensive platinum was ruined for nothing. (Fortunately for Moissan, he had a rich father-in-law, who subsidized him generously.)

Moissan, like Frémy before him, decided to back away from straight chemistry and try electrolysis.

He began with arsenic fluoride and after fiddling with that, unsuccessfully, he decided to abandon that line of investigation because he was beginning to suffer from arsenic poisoning. So he turned to hydrogen fluoride (and underwent four different episodes of hydrogen fluoride poisoning, which eventually helped bring him to his death at the age of fifty-four).

Moissan remembered perfectly well that Frémy's anhydrous hydrogen fluoride would not carry an electric current. Something had to be added to make it do so, but not something that would offer an alternate element for production at the positive electrode. Why not another fluoride? Moissan dissolved potassium hydrogen fluoride in the anhydrous hydrogen fluoride and had a mixture which could pass a current and which could produce only fluorine at the positive electrode.

Furthermore, he made use of equipment built up out of an alloy of platinum and iridium, an alloy that was even more resistant to fluorine than platinum itself was.

Finally, he brought his entire apparatus to $-50°$ C. All chemical reactions are slowed as temperature decreases and at $-50°$ C even fluorine's savagery ought to be subdued.

Moissan turned on the current and hydrogen bubbled off the negative electrode like fury, but nothing showed at the positive electrode. He stopped to think. The positive electrode was inserted into the platinum-iridium vessel through a stopper. The stopper had to be an insulator so it couldn't be platinum or any metal; and that stopper had been eaten up by fluorine. No wonder he hadn't gotten any gas.

Moissan needed a stopper made of something that would not carry an electric current and would be untouched by fluorine. It occurred to him that the mineral fluorite already had all the fluorine it could carry and would not be attacked further. He therefore carefully carved stoppers out of fluorite and repeated the experiment.

On June 26, 1886, he obtained a pale yellow-green gas about his positive electrode. Fluorine had finally been isolated, and when Moissan later repeated the experiment in public, his old teacher Frémy watched.

Moissan went on, in 1899, to discover a less expensive way of producing fluorine. He made use of copper vessels. Fluorine attacked copper violently, but after the copper was overlaid with copper fluoride, no further attacks need be expected. In 1906, the year before his death, Moissan received the Nobel Prize in chemistry for his feat.

Even so, fluorine remained the bad boy of the table of elements for another generation. It could be isolated and used; but not easily and not often. Most of all, it couldn't be handled with anything but supreme caution for it was even more poisonous than hydrogen fluoride.

Meanwhile the noble gases were discovered in the 1890s (see Chapter 4) and although they were recognized as being extremely inert, chemists tried over and over to force them into some kind of compound formation. (Don't believe the myth that chemists were so sure that the noble gases wouldn't react that they never tried to test the fact. Dozens of compounds were reported in the literature—but the reports, until quite recently, always proved to be mistaken.)

It wasn't until the early 1930s that chemical theory had been developed to the point where one need not tackle the noble gases at random in an effort to form compounds. The American chemist Linus Pauling, in 1933, was able to show, through logical arguments, that xenon ought to be able to form compounds with fluorine. Almost at once two chemists at Pauling's school, the California Institute of Technology, took up the challenge. They were Donald M. Yost and Albert L. Kaye.

All the xenon they could get hold of was 100 cc worth at normal air pressure and they could get hold of no fluorine. They had to rig up a device of their own to prepare fluorine; and it worked only intermittently. Doing the best they could, they found they could obtain no clear signs of any compound. Neither were they completely certain that no compound had been formed. The results were inconclusive.

There was no immediate follow-up. The results didn't warrant it. Chemists knew the murderous history of fluorine, and enthusiasm for such experiments ran low.

During World War II, fluorine was needed in connection with atomic bomb research. Under that kind of pressure, methods for the production of fluorine in quantity, and *in reasonable security* were developed.

By the 1950s, it was finally possible to run nonmilitary experiments, involving fluorine, without much risk of suicide. Even then, there were only a few laboratories equipped for such work and those had a great many things to do with fluorine other than mixing them with noble gases.

"Just mix xenon and fluoride in a nickel container" indeed. It could not have been done, in reasonable safety and with reasonable hopes of success, any more than ten years before it actually was done in 1962; and, under the circumstances, the ten years' delay was a remarkably reasonable one and reflects no discredit whatever on Science.

6

TO TELL A CHEMIST

Some time ago, I watched a television program called "To Tell the Truth." If you are unaware of its nature, I will explain that it involves a panel of four, who try to guess which one, of three people claiming to be John Smith, is the *real* John Smith. They do so by asking questions which, they hope, the real John Smith (pledged to tell the truth) can answer correctly, while the phonies cannot.

The reason I watched was that Catherine de Camp (the lovely and charming wife of L. Sprague de Camp) was scheduled to appear as a contestant in her capacity as archaeologist. To my surprise, two of the four panelists would not believe she was the real Catherine de Camp. Her case seemed shaken when, in answer to one question, she stated that Atlantis had never existed.[1] The stir of disapproval among the panelists was marked. Surely, no real archaeologist (they were plainly thinking) would deny the existence of Atlantis.

[1] *Since this article was written in 1965, it has been discovered that the small island of Thera in the Aegean Sea exploded volcanically in 1400 B.C. and that it may have inspired the legend of Atlantis, but the legend described a place that had changed from the original out of all recognition.*

And it got me to thinking—

How does one distinguish quickly and easily between a specialist and a well-primed nonspecialist? It seems to me you must find little things no one would ever think to prime the nonspecialist upon.

Since I know the chemical profession best, I devised two questions, for instance, to tell a chemist from a nonchemist. Here they are:

(1) How do you pronounce UNIONIZED?

(2) What is a mole?

In response to the first question, the nonchemist is bound to say "YOON-yun-ized," which is the logical pronunciation, and the dictionary pronunciation, too. The chemist, however, would never think of such a thing; he would say without a moment's hesitation: "un-EYE-on-ized."

In response to the second question, the nonchemist is bound to say, "A little furry animal that burrows underground," unless he is a civil engineer who will say, "A breakwater." A chemist, on the other hand, will clear his throat, and say, "Well, it's like this—" and keep talking for hours.

There's my cue. Shall we talk about the chemical version of the little furry animal?

To do so, we will begin with molecules. The oxygen molecule, consisting of 2 oxygen atoms, has a molecular weight of 32; while the hydrogen molecule, consisting of 2 hydrogen atoms, has a molecular weight of 2. Such molecular weights are pure numbers and it is not necessary to go into their significance here. All we have to understand at this moment is that the ratio of the mass of an oxygen molecule to that of a hydrogen molecule is indicated by their respective molecular weights to be 32 to 2.

If we take 2 molecules of oxygen and 2 of hydrogen, the mass of each substance is doubled, but the ratio remains the same. The ratio also remains the same if we take 10 of each type of molecule, or 100 of each, or 5,266 of each, and so on.

We can make it general and say that as long as we have equal numbers of molecules of hydrogen and of oxygen, the total mass of the oxygen molecules is to the total mass of the hydrogen molecules as 32 is to 2.

We can begin with a 2-gram sample of hydrogen. This contains a certain number of hydrogen molecules, which we will call N. Imagine that we also have a sample of oxygen which contains N oxygen molecules. Since the two gas samples contain equal numbers of molecules, the mass of the oxygen to that of the hydrogen is as 32 is to 2. The mass of the hydrogen has been set at 2 grams, therefore the mass of the oxygen is 32 grams.

We conclude that 2 grams of hydrogen and 32 grams of oxygen both contain N molecules.

Notice the significance of the 2-gram sample of hydrogen. It is the numerical value of the molecular weight (2) expressed in grams. We can therefore refer to 2 grams as the "gram-molecular-weight" of hydrogen. (Similarly, 2 pounds of hydrogen would be the pound-molecular-weight, 2 tons of hydrogen would be the ton-molecular-weight, and so on. We will confine ourselves, however, to gram-molecular-weights.)

By the same reasoning, 32 grams of oxygen is a gram-molecular-weight of oxygen.

Now the phrase "gram-molecular-weight" contains six syllables. Since chemists must use the phrase very frequently, they sought avidly for some shortened version. You will note that the fifth to eighth letters inclusive are "m-o-l-e." With a wild cry of delight, chemists shortened "gram-molecular-weight" to "mole."

Some of them, in the nervous realization that a "mole" is a little, furry animal that burrows underground, try to use "mol" instead. I was forced to use "mol," in a textbook I once wrote, by the overriding vote of my two coauthors, a state of affairs which led to internal bleeding. The word is universally pronounced with a long "o" and "mol" must clearly have a short "o." Consequently, in this chapter, where I am my own master, I use "mole." Do you hear me, world? "Mole!"

Very well, then, I have already shown that 1 mole of hydrogen and 1 mole of oxygen both have the same number (N) of molecules. By similar reasoning, it is possible to show that 1 mole of any substance at all contains N molecules.

As examples, the molecular weight of water is 18, that of sulfuric acid is 98, and that of table sugar (sucrose) is 342. There are, therefore, N molecules in 18 grams of water, in 98 grams of sulfuric acid, and in 342 grams of sucrose.

Now I have explained the mole, but one thing leads to another, and I refuse to stop.

For instance, suppose you collect 1 mole of hydrogen (2 grams) and keep it at what is called "standard temperature and pressure" (STP), which means a temperature of 0° C and a pressure of 1 atmosphere. You will find that the hydrogen will take up a volume of 22.4 liters.

Suppose you next do the same for 1 mole of oxygen (32 grams). Its volume at STP is *also* 22.4 liters. In fact, take 22.4 liters of any gas, and though the mass of the gas may vary all over the lot, you will always find yourself with 1 mole.[2]

In the same way, 11.2 liters of any gas contain 0.5 moles of that gas; 44.8 liters of any gas contain 2 moles of that gas; and so on. In fact, we can make the following statement: "Equal volumes of gases under fixed conditions of temperature and pressure contain equal numbers of molecules."

This statement is easy to work out once there is a grasp of the atomic theory of matter, plus the simple observation that 2 grams of hydrogen and 32 grams of oxygen take up the same volume.

The statement was first made in 1811, however, by an Italian physicist named Amedeo Avogadro (see Chapter 1), at a time when the atomic theory had just been broached and was barely invading the chemical consciousness. The statement (still called "Avogadro's hypothesis" to this day) seemed, at the time it was made, to be pulled out of thin air and was generally ignored. It took fifty years before its worth and value were appreciated and, as you might expect, Avogadro died just a few years too soon to see himself vindicated.

The next question is, what is the value of N? How many molecules are there in 1 mole of any substance? Obviously, it is a very large number since molecules are so small, but that was as far as anyone could go at first. Avogadro, in his lifetime, hadn't the slightest idea of what the exact value of N might be; and neither had anyone else.

It wasn't until 1865 that a German physicist, J. Loschmidt, worked out a reasonable value for the first time, following a particular theoretical approach. Since then, at least a dozen different approaches have been utilized, and all have yielded virtually the same result. The number of molecules in 1 mole of a substance (called "Avogadro's number," by the way) turns out to be, using the value officially accepted in 1963, 6.02252×10^{23}. If you want that written out in full, it is 602,252,000,000,-000,000,000,000; or, in words, it is a little over six hundred sextillion.

From Avogadro's number, you can work out the actual mass of any

[2] *Actually, this is precisely true only in the case of a "perfect gas," which I will mention again later in the article. Actual gases deviate slightly from this state of affairs, and some gases deviate quite a bit. To make my point here, however, I shall overlook minor imperfections.*

molecule, by dividing the number into the molecular weight. Thus, since 32 grams of oxygen contains 6.02252×10^{23} oxygen molecules, one oxygen molecule has a mass of 32 divided by 6.02252×10^{23}, or about 5.31×10^{-23} grams (0.0000000000000000000000531 grams).

It may seem unfair to you that Avogadro's name is attached to a number he never worked out, but it doesn't to me, for he was the one who made the crucial mental leap in this respect. However, if you are one who finds the apparent unfairness rankling, feel relieved! Loschmidt, who first worked out the value of Avogadro's number, is himself appropriately honored. The number of molecules in 1 cubic centimeter of gas at STP is "Loschmidt's number." Since 1 mole of gas takes up 22.4 liters, or to be more precise, 22,415 cubic centimeters, at STP, Loschmidt's number is Avogadro's number divided by 22,415.

Loschmidt's number therefore comes out to be 2.68683×10^{19}, or 26,868,300,000,000,000,000, or just under twenty-seven quintillion.

Now we can have fun and games with Loschmidt's number (which we will symbolize as L).

If there are L molecules in 1 cubic centimeter of gas, then the average distance between the center of one molecule and that of its neighbor is equal to the reciprocal of the cube root of L; that is to $1/\sqrt[3]{L}$.

Working this out (I'll do it myself; I needn't plague you with everything), it becomes apparent that the average intermolecular distance in a gas at STP is 3.33×10^{-7} centimeters. This is a very short distance for it is about a third of a millionth of a centimeter and a centimeter is about two-fifths of an inch. We might well feel justified in considering gases to be choked to bursting with molecules.

Let's consider matters further, however. A hundred-millionth of a centimeter (10^{-8} centimeters) is an "Ångstrom unit," which is usually abbreviated as Å. This means that the average intermolecular distance in a gas at STP can be expressed as 33.3 Å.

But the radius of a small molecule is in the neighborhood of a little over 3 Å. This means that the separation between small molecules is some 10 times the radius of those same molecules. If one of those molecules were expanded to the size of the Earth, its neighbor (also the size of the Earth) would be 40,000 miles away, or something more than one-sixth the distance between the Earth and the Moon. That might be quite close astronomically, but certainly the Earth would not feel particularly crowded with a neighbor at such a distance.

In fact, the amount of space taken up by small gas molecules would be only $\frac{1}{1000}$ of the total volume of the gas. To put it another way, ordinary gases are something like 99.9 percent intermolecular space and only 0.1 percent molecules.

From that standpoint, gases aren't crowded with matter at all. They might, instead, be looked upon as reasonable approaches to vacuum.

Notice that I've been specifying standard temperature and pressure. If the pressure is increased it is easy to push the molecules closer together, considering how much empty space there is in gases. In fact, doubling the pressure halves the volume of the gas, tripling the pressure reduces the volume to one-third, and so on (provided there is no temperature change).

You might wonder why the gas molecules don't fall together of their own accord. Why should they stay so far apart anyway? The answer is that they possess energy which expresses itself in the form of rapid motion, and this motion jostles the molecules apart, so to speak, through incessant collisions. If the pressure is relieved, the molecular jostling moves the molecules correspondingly farther apart. If the pressure is reduced to one-half, the volume of the gas doubles; if the pressure is reduced to one-third, the volume triples, and so on (provided, again, there is no temperature change).

If the temperature is increased (and the pressure is left unchanged), the molecular velocity increases, the jostling is more energetic and the volume increases. If the temperature falls, the volume decreases. There is thus a neat interlocking among the temperature, pressure, and volume of a particular sample of gas. If the gas is perfect, the relationship can be expressed as a very simple "equation of state." For actual gases, the equation has to be modified and made more complicated, but we'll discuss that another time, perhaps.

The first to note the relationship of pressure and volume in gases was the English chemist Robert Boyle, in 1662. In 1677, a French physicist, Edme Mariotte, discovered the relationship independently and was the first to specify that temperature must be kept unchanged. In Great Britain and America, we therefore speak of "Boyle's Law" and in Continental Europe of "Mariotte's Law."

In 1699, a French physicist, Guillaume Amontons, noted the effect of temperature on air, and the manner in which volume and temperature were interrelated. Another French physicist, Jacques A. C. Charles, repeated the observation in 1787 and noted that it applied to all gases and not to air alone. Charles did not publish, however, and a French chemist, Joseph Louis Gay-Lussac, who repeated the observation yet again in 1802, *did* publish. The relationship is therefore referred to as either "Charles's law" or "Gay-Lussac's law." Poor Amontons gets nothing.

So far, the development of understanding concerning the equation of state for gases was the result of purely empirical observation. In the 1860s, however, the Scottish mathematical physicist James Clerk Maxwell accepted a gas as a collection of perfectly elastic molecules engaged in rapid random motion and treated the collection of molecules by means of a rigorous statistical interpretation. An Austrian physicist, Ludwig Boltzmann, did the same independently. Together, they showed that such an interpretation could account for the pressure/temperature/volume relationships beautifully.

Thus was developed the "kinetic theory of gases" ("kinetic" coming from a Greek word for "motion") and it was from this kinetic theory, and the equations it produced, that Loschmidt worked out Avogadro's number for the first time. See how science hangs together!

Maxwell's kinetic theory made use of two assumptions that aren't perfectly correct. To simplify matters, he supposed that the individual gas molecules were of zero size and that there was no mutual molecular attraction. A gas for which these assumptions are correct is the perfect gas I mentioned earlier. In actual gases, the molecules are tiny, but not of zero size, and there is a tiny, but not zero, mutual attraction. Hence, actual gases are more or less imperfect. The imperfection is least in the cases of the gases helium, hydrogen, and neon, where the molecules (or, in the case of helium and neon, single atoms) are smallest and the mutual attraction least.

We can pretend, though, that we are dealing with a perfect gas and consider the effect of temperature. If we begin with a mole of perfect gas at STP, we find the volume is 22,415 cubic centimeters. For every degree C we raise the temperature, the volume increases by a trifle over 82 cubic centimeters, and for every degree C by which we drop the temperature, the volume decreases by a trifle over 82 cubic centimeters.

If we continue dropping the temperature, degree by degree, and if 82 cubic centimeters peels off the volume with each degree, then by the time we reach a temperature of $-273.15°$ C, the volume has decreased to zero. It was this fact which first gave rise to the notion of $-273.15°$ C as an "absolute zero," an ultimate cold which could not be surpassed.

Of course, it is only in a perfect gas with molecules of zero size that a shrinkage of volume to zero can be visualized. In any actual gas, with molecules of some definite size, volume can shrink only to the point where the molecules make surface-to-surface contact, at which point the situation changes radically.

Suppose that the molecules of a particular gas have a radius of 1 Å. At surface contact, the molecules are separated, center to center, by a distance equal to the sum of their radii; that is, by a distance of 2 Å. We can calculate at what temperature this should happen.

At $0°$ C the center-to-center separation is 33.3 Å and at $-273.15°$ C the separation is (ideally) zero. The distance declines smoothly with falling temperature so that we find that at $-257°$ C the separation has decreased to 2 Å and surface-to-surface contact has been made. Since $-257°$ C is about $16°$ above absolute zero, it can be written $16°$ K (where K stands for Kelvin; Lord Kelvin having been the first to make use of a temperature scale that placed zero at the absolute zero).

If the molecule is particularly small so that the radius is only 0.5 Å, surface-to-surface contact would be made at a temperature of $8°$ K.

Once surface-to-surface contact is made, the substance—under ordinary circumstances at least—is not likely to behave as a gas any further. We have, instead, a "condensed phase."

When surface-to-surface contact is first made, the molecules will still possess sufficient energy to slide around freely. They are then in the "liquid state." If the temperature falls lower and energy is further subtracted, the molecules lock into place and the substance is in the "solid state."

It would seem from what I have said so far that the perfect gas would never liquefy since its molecules would never make surface-to-surface contact, short of absolute zero itself, and absolute zero cannot be reached. Actual gases, however—or so it would seem—must liquefy at temperatures short of absolute zero, but not very far short.

This is more or less true for the three actual gases which, of all gases, are nearest to perfection. Helium, the most nearly perfect, liquefies at $4.2°$ K, hydrogen at $20.3°$ K, and neon at $27.2°$ K. Other gases, however, liquefy at considerably higher temperatures. Oxygen, for instance, which is not terribly imperfect, has a liquefaction point of $90.1°$ K.

At 90.1° K, the molecules of gas have an average separation, center-to-center, of about 11 Å. Even if we allow the oxygen molecule a radius of 2 Å, the surface-to-surface separation would be 7 Å. The temperature could drop down to close to 30° K before surface-to-surface contact was made.

Nevertheless, oxygen liquefies at 90.1° K and not at 30° K. To explain that, we must remember the second imperfection of actual gases; the fact that there is an attraction between molecules. In the case of helium, hydrogen, and neon, this attraction is very small. If helium atoms happen to collide the mutual attraction is so small it is easily overcome even by the small amount of energy of motion present at extremely low temperatures. For that reason, liquefaction of helium doesn't take place till surface-to-surface contact enforces it.

The mutual attraction among oxygen molecules, however, is considerably higher than among helium or neon atoms or among hydrogen molecules. By the time the temperature has sunk to 90.1° K, the energy of motion is no longer sufficient to pull apart two molecules that have happened to collide. The attraction among oxygen molecules is sufficiently large to hold the combination in place, and oxygen liquefies.

A great many substances possess intermolecular (or interatomic or interionic) attractions so great that they are not gases even at high temperatures; a few not until a temperature of 6000° C is reached.

Now let's tackle the condensed phases, beginning with liquid hydrogen. This has a density of 0.07 grams per cubic centimeter at its boiling point (the lowest density, in fact, for the condensed phase of any substance).

Since 2 grams of hydrogen (1 mole) contain 6.02252×10^{23} molecules, 0.07 grams contain approximately 2.09×10^{22} molecules. The average center-to-center separation of the molecules is, therefore, 3.63 Å. This can be taken as the effective diameter of the hydrogen molecule in liquid hydrogen. (For an oxygen molecule, similar calculations yield a diameter of about 3.9 Å.)

You might suppose, that as one went up the table of elements to more and more complex atoms, that the atomic diameters, calculated from the density of the condensed phases of the elements, would get steadily larger. This, however, is not so.

The atomic volume is largely determined by the amount of space taken up by the electrons of the atom, and a great deal depends on just how those electrons are arranged. The electrons are arranged in shells and in some atoms, the outermost shell is occupied by a single electron, which is usually held quite weakly and moves far out from the nucleus, giving that atom an unusually large volume.

This is true for sodium, potassium, rubidium, and cesium, for instance, with cesium the most extreme case, for it has more electrons all together than the other atoms of its type.

Cesium, like metals generally, is considered as being made up of single atoms not arranged in molecular combinations. The atomic weight of cesium is 132.9 so that 132.9 grams is the "gram-atomic-weight." (That is not a gram-molecular-weight, so it shouldn't, strictly speaking, be referred to as a "mole.") The gram-atomic-weight of an element contains Avogadro's number of atoms.

The density of cesium at room temperature is 1.87 grams per cubic centimeter so that 1 cubic centimeter of cesium contains about 8.15×10^{21} atoms. The effective diameter of the cesium atom in solid cesium is therefore about 5 Å.

On the other hand, when the outermost shell is about half full of electrons, the atom is quite small. The electrons are drawn unusually

DMITRI IVANOVICH MENDELEEV

Mendeleev, born in Tobolsk, Siberia, on February 7, 1834, came of a large family in which there were between fourteen and seventeen children, the records not being exactly clear. Dmitri was the youngest and he probably had some Asian ancestry, for his mother is supposed to have been part Mongol. Mendeleev received his first lessons in science from a political prisoner who had been sent to Siberia.

In 1849, when Mendeleev was just finishing high school, his father died and his mother's glass factory burned down, leaving her without means. Most of her children had become independent, so Mendeleev's mother took him first to Moscow and then to St. Petersburg. She managed to get the young man into college and then she died.

Mendeleev finished college in 1855 at the top of his class, went to France and Germany for graduate training, then returned to St. Petersburg to become a professor of chemistry at the university there. He was the most capable and interesting lecturer in Russia and one of the best in all Europe. It was through him that it finally became possible for young Russians to receive graduate training in chemistry

without leaving Russia. Between 1868 and 1870 he wrote a chemistry textbook that was probably the best ever written in Russian. It had numerous footnotes that took up almost as much space as the book itself.

Mendeleev worked out the periodic table of the elements in 1869, a feat which rationalized the study of the elements at last. To make the table work, he did what no other had dared do; he left blanks which he said were to be filled by elements yet undiscovered. To Western chemists, this sounded like typical Russian mysticism. Within fifteen years, however, three elements were found which in all respects proved to have properties matching those predicted by Mendeleev, who suddenly became the most famous chemist in the world. A liberal who was a thorn in the side of the despotic Russian government, he was nevertheless sent to the United States to study oil technology. He died in St. Petersburg, on February 2, 1907.

SVANTE AUGUST ARRHENIUS

Arrhenius was born at Wijk, Sweden, on February 19, 1859, and was an infant prodigy, teaching himself to read at three. He graduated from high school as the youngest and brightest in his class.

While attending the University of Uppsala, he began to study how electricity passed through solutions, and it seemed to him that the puzzles involved in the study could best be explained if one supposed that atoms or molecules could break up into electrically charged fragments called "ions."

For his Ph.D. thesis in 1884 he prepared a carefully worked out presentation of his "ionic theory" but it turned out to be a bit too revolutionary as a concept. His examining professors were firmly convinced that atoms could not break up and could not bear an electric charge. He just barely passed.

Fortunately, a few of the younger men in the new field of physical chemistry were intrigued by the notion and by the young man's close to the central nucleus, and this means that neighboring atoms can be drawn unusually close together.

In fact, the compactness of packing proceeds in periodic waves if one plots it against atomic weight. The atomic diameter rises to a peak, and packing is least compact each time a one-electron-in-the-outermost-shell point is reached; and atomic diameter falls to a trough, and packing is most compact, each time the outermost-shell-half-full situation is reached. It was this which, in 1870, gave the German chemist Lothar Meyer the notion of the "periodic table" of elements. (Meyer, however, was beaten to the punch by the Russian chemist Dmitri I. Mendeleev, who reached the same conclusion by another line of argument just a few months earlier. But that is another story.)

Examples of regions in the periodic table of particularly small atoms are, in order of increasing complexity of atomic structure: (1) beryllium, boron, and carbon, (2) iron, cobalt, and nickel, (3) ruthenium, rhodium, and palladium, and (4) osmium, iridium, and platinum.

Without going into all the mathematical details, here are some interatomic distances in the room-temperature solid (and, therefore, the effective atomic diameters). Carbon (in the form of diamond) 1.8 Å; nickel, 2.2 Å; rhodium, 2.4 Å; and osmium 2.4 Å.

Diamond is the most compact of all solids. This, combined with the fact that each carbon atom in diamond is firmly held by each of four

reasoning. Arrhenius worked with them and his reputation increased as he evolved another theory of "energy of activation," which was essential to the theory of chemical reactions.

In the 1890s, when electrically charged subatomic particles were discovered, Arrhenius' ionic theory suddenly made sense and in 1903 he received the Nobel Prize in chemistry for it.

In 1908 Arrhenius advanced a new concept that had less success but has remained interesting to scientists and science fiction readers alike. He suggested that life might have originated on worlds through spores blown up through the atmosphere of life-bearing worlds and then driven by radiation pressure across the gulfs of space to fall on hospitable but as yet sterile worlds. He also pointed out the existence of a "greenhouse effect" in which small changes in the concentration of carbon dioxide in the atmosphere could considerably alter the average temperature of a planet. He died in Stockholm on October 2, 1927.

record-close neighbors, is what makes diamond the hardest known substance (with the possible exception of boron nitride, which closely mimics the diamond situation).

The more compact a solid is, the denser it is, and the more massive the individual atoms are, the more extreme the density. Of the various groups of compact atoms, the most massive are those of the three elements, osmium, iridium, and platinum. They should be, therefore, and *are*, the densest of the elements (or, indeed, of any substance).

The density of platinum is 21.37 grams per cubic centimeter, that of iridium is 22.42 grams per cubic centimeter, and that of osmium, the record holder, 22.5 grams per cubic centimeter. Osmium is just about twice as dense as lead, and is $\frac{1}{6}$ denser than gold. A cubic foot is not a very large volume, but a cubic foot of osmium weighs 1,400 pounds.

Naturally, the farther apart atoms are (center-to-center) the less trouble it is, all other things being equal, to pull them apart altogether, whether by heat or by the chemical pull of other atoms. Thus, the loosely packed cesium has a melting point of 28.5° C and a boiling point of 670° C, while osmium melts at 2,700° C and boils at some temperature higher than 5,300° C.

Of all solids, carbon is the most compact, and it also has the highest melting point. It is close to 3,700° C before it ceases to be a solid. (Actually it sublimes, rather than melts, turning into gaseous carbon.)

Again, cesium is so ready to leave the society of its fellows and join with other atoms that it is the most active of all metals. Osmium, iridium, and platinum, are, on the other hand, the least active of all metals.

You see?

Beginning students of chemistry often think of the science as a mere collection of disconnected data to be memorized by brute force. Not at all! Just look at it properly and everything hangs together and makes sense.

Of course, getting the hang of the proper look isn't always easy.

PART II

NUCLEAR CHEMISTRY

7

THE EVENS HAVE IT

Some time ago, I was asked (by phone) to write an article on the use of radioisotopes in industry. The gentleman doing the asking waxed enthusiastic on the importance of isotopes, but after a while I could stand it no more, for he kept pronouncing it ISS-o-topes, with a very short "i."

Finally, I said, in the most diffident manner I could muster, "EYE-so-topes, sir," giving it a very long "i."

"No, no," he said impatiently, "I'm talking about ISS-o-topes."

And so he did, to the very end, and on subsequent phone calls too. But I fooled him. I eventually wrote the article about EYE-so-topes.

Yet it left a sore spot, for having agreed to do the article, I was forced to deal with only the practical applications of isotopes, a necessity which saddened me. There is much that is impractical about isotopes that I would like to discuss, and I will do so here.

The way in which "isotope" came into the scientific vocabulary is a little involved. A.ter two millennia of efforts, most of the elements making up the universe had been isolated and identified. In 1869, the Rus-

ERNEST RUTHERFORD

Rutherford's grandfather was a Scotsman who had emigrated to New Zealand in 1842. His father was a farmer, and Rutherford, born near Nelson, New Zealand, on August 30, 1871, as the second of twelve children, worked on the farm. In 1895 Rutherford placed second in a competition for a scholarship to Cambridge. The first-place winner wanted to stay in New Zealand and get married, so Rutherford received his chance. The news reached him while he was digging potatoes on his father's farm. He flung down his spade and never dug another potato. He postponed his own marriage plans to go to England.

sian chemist Dmitri Ivanovich Mendeleev arranged the known elements in order of atomic weights and showed that a table could be prepared in which the elements, in this order, could be so placed as to make those with similar properties fall into neat columns.

By 1900, this "periodic table" was a deified adjunct of chemistry. Each element had its place in the table; almost each place had its element. To be sure, there were a few places without elements; but that bothered no one since everyone knew that the list of known elements was incomplete. Eventually, chemists felt certain, an element would be discovered for every empty place in the table. And they were right. The last hole was filled in 1948, and additional elements were discovered beyond the last known to Mendeleev. As of now, 103 different elements are known.[1]

After 1900, however, the much more serious converse of the situation

[1] *This article was written in 1961. Since then, two more elements, with atomic numbers 104 and 105, have been reported.*

There he began work in the exciting new field of radioactivity. He studied the different radiations and named them "alpha," "beta," and "gamma." He also studied the rate at which radioactive atoms broke down and coined the term "half-life."

Between 1906 and 1909 he studied alpha particles intensively and proved they were the nuclei of helium atoms. In 1914 he suggested that the hydrogen nucleus was the simplest positively charged particle and named it the "proton."

He also studied the manner in which alpha particles behaved when striking a thin sheet of metal foil. He found that for the most part the particles went through undisturbed and undeflected, but every once in a while a particle bounced. He maintained in 1911 that this occasional bounce showed the atoms in the foil were made up, for the most part, of light atoms and that only a tiny region in the center contained the massive protons. Thus was developed the notion of the "nuclear atom." In 1917 Rutherford showed that under the bombardment of atoms by subatomic particles, nuclei could be changed in structure. He thus changed one element into another and achieved transmutation—though not in any form the ancient alchemists dreamed of.

He received the Nobel Prize in chemistry in 1908 and was created Baron Rutherford in 1931. After 1933 he was strongly anti-Nazi in his sympathies and helped many Jewish scientists who had been forced to flee Germany. He died in Cambridge on October 19, 1937.

arose. Substances were found among the radioactive breakdown products of uranium and thorium which had to be classified as new elements by nineteenth-century standards, since they had properties unlike those of any other elements—and yet there was no place for them in the periodic table.

Eventually several scientists, notably the British physicist Frederick Soddy, swallowed hard and decided that it was possible for two or more elements to occupy the same place in the periodic table. In 1913, Soddy suggested the name "isotope" for such elements, from Greek words meaning "same place."

An explanation rehabilitating the periodic table followed in due course. The New Zealand-born British physicist Ernest Rutherford had already (in 1906) shown that the atom consisted of a tiny central nucleus containing positively-charged protons and of a comparatively vast outer region in which negatively-charged electrons whirled. The number of protons at the center is equal to the number of electrons in the outskirts, and since the size of the positive electric charge on a proton (arbitrarily set at $+1$) is exactly equal to the size of the negative electric charge on an electron (which is, naturally, -1), the atom as a whole is electrically neutral.

The next step was taken by a young English physicist named Henry Gwyn Jeffreys Moseley. By studying the wave lengths of the X rays emitted, under certain conditions, by various elements, he was able to deduce that the total positive charge on the nucleus of each element had a characteristic value. This is called the "atomic number."

For instance, the chromium atom has a nucleus with a positive charge of 24, the manganese atom one of 25 and the iron atom one of 26. We can say then that the atomic numbers of these elements are 24, 25, and 26 respectively. Furthermore, since the positive charge is entirely due to the proton content of the nucleus, we can say that these three elements have 24, 25, and 26 protons in their nuclei, respectively, and that circling these nuclei are 24, 25, and 26 electrons, respectively.

Now throughout the nineteenth century it had been held that all atoms of an element were identical. This was only an assumption, but it was the easiest way of explaining the fact that all samples of an element had identical chemical properties and identical atomic weights.

But this was when atoms were viewed as hard, indivisible, featureless spheres. How did the situation stand up against the twentieth-century notion that the atoms were complex collections of smaller particles?

X-ray data showed that the atomic number of an element was a matter of absolute uniformity. All atoms of a particular element had the same number of protons in the nucleus and therefore the same number of electrons in the outskirts. Through the 1920s it was shown that the chemical properties of a particular element depended on the number of electrons it contained and that therefore all atoms of an element had identical chemical properties. Very good, so far.

The matter of atomic weight was not so straightforward. To begin with, it was known from the first days of nuclear-atomic theory that the nucleus must contain something other than protons. For instance: the nucleus of the hydrogen atom was the lightest known and it had a positive charge of 1. Consequently it seemed quite reasonable, and even inevitable, to suppose that the hydrogen nucleus was made up of a single proton. Its atomic weight, which had been set equal to 1 (not quite, but just about) long before the days when atomic structure had been worked out, turned out to make sense.

Helium, on the other hand, had an atomic weight of 4. That is, its nucleus was known to be four times as massive as the hydrogen nucleus. The natural conclusion seemed to be that it must contain four protons. However, its atomic number, representing the positive charge of its nucleus, was only 2. An equally natural conclusion from that seemed to be that the nucleus must contain only two protons.

With two different but natural conclusions, something had to be done. The only other subatomic particle known in the first decades of the twentieth century was the electron. Suppose then that the helium nucleus contained four protons and two electrons. The atomic weight would be 4 because the electrons weigh practically nothing. The atomic number, however, would be 2 because the charge on two of the protons would be canceled by the charge on the two electrons.

There were difficulties in this picture of the nucleus, however. For instance, it gave the helium nucleus six separate particles, four protons and two electrons, and that didn't fit in with certain other data that were being accumulated. Physicists went about biting their nails and talking in low, glum voices.

Then, in 1932, the neutron was discovered by the English physicist James Chadwick, and it turned out that all was right with the theory, after all. The neutron is equal in mass to the proton (just about), but has no charge at all. Now the helium nucleus could be viewed as consisting of two protons and two neutrons, you see. The positive charge and hence the atomic number would be 2 and the atomic weight would be 4. This would involve a total number of four particles in the helium nucleus, and that fit all data.

Now, how does the presence of neutrons in the nucleus of an atom affect the chemical properties? Answer: It doesn't; at least, not noticeably.

Take as an example the copper atom. It has an atomic number of 29, so every copper atom has twenty-nine protons in the nucleus and twenty-nine electrons in the outer reaches. But copper has an atomic weight of (roughly) 63, so the nucleus of the copper atom must contain, in addition to twenty-nine protons, thirty-four neutrons as well. The neutrons have no charge; they do not need to be balanced. The twenty-nine electrons balance the twenty-nine protons and, as far as they are concerned, the neutrons can go jump in the lake.

Well, then, suppose just for fun that a copper atom happened to exist with a nucleus containing twenty-nine protons and thirty-six neutrons, two more neutrons, that is, than the number suggested in the previous paragraph. Such a nucleus would still require only twenty-nine electrons to balance the nuclear charge; and the chemical properties, which depend on the electrons only, would remain the same.

In other words, if we judge by chemical properties alone, the atoms of an element need *not* be identical. The number of neutrons in the nucleus could vary all over the lot and this would make no difference chemically. Since the periodic table points out chemical similarities and since the elements are defined by their chemical properties, it means that each place in the periodic table is capable of holding a large variety of different atoms, with different numbers of neutrons, *provided* the number of protons in all those atoms is held constant.

But how does this affect the atomic weight?

The two varieties of copper atoms would, naturally, be well mixed at all times. Why not? Since they would have identical chemical properties, they would travel the same path in geochemical processes; all of them would react equally with the environment about them, go into solution and out of solution at the same time and to the same extent. They would be inseparable; in the end, any sample of an element found in nature, or prepared in the laboratory, would contain the same even mixture of the two copper isotopes.

In obtaining the atomic weight of an element, then, nineteenth-century chemists were getting the *average* weight of the atoms of that element (see Chapter 1). The average would always be the same (for anything *they* could do), but that did *not* mean that all the atoms were individually identical.

Then what happened to upset this comfortable picture once radioactivity was discovered?

Well, radioactive breakdown is a *nuclear* process, and whether it takes place or not, and how quickly, and in what fashion, depends on the arrangement of particles in the nucleus and has nothing to do with the electrons outside the nucleus. It follows that two atoms with nuclei containing the same number of protons but different numbers of neutrons would have identical chemical properties but different nuclear properties. It was the identical chemical properties that placed them in the same spot in the periodic table. The different nuclear properties had nothing to do with the periodic table.

But in the first decade of the twentieth century, when the distinction between nuclear properties and chemical properties had not yet been made, there was this period of panic when it seemed that the periodic table would go crashing.

It was easy to distinguish between two isotopes (which, you now see, are two atoms with equal numbers of protons in their nuclei but different numbers of neutrons) if radioactivity was involved. What, however, if neither of two isotopes is radioactive? Is it even possible for there to be more than one nonradioactive isotope of a given element?

Well, if a plurality of nonradioactive isotopes of an element existed, they would differ in mass. A copper atom with 29 protons and 34 neutrons would have a "mass number" of 63, while one with 29 protons and 36 neutrons would have one of 65. (The expression "atomic weight" is reserved for the average masses of naturally occurring mixtures of isotopes of a particular element.)

In 1919, the English physicist Francis William Aston invented the mass spectrograph in which atoms in ionic form (that is, with one or more electrons knocked off so that each atom has a net positive charge) could be driven through a magnetic field. The ions follow a curved path in so doing, the sharpness of the curve depending on the mass of the particular ion. Isotopes having different masses end on different spots of a photographic plate, and from the intensities of darkening, the relative quantities of the individual isotopes can be determined. For instance, the 34-neutron copper atom makes up 70 per cent of all copper atoms while the 36-neutron copper atom makes up the remaining 30 per cent. This accounts for the fact that the atomic weight of copper is not exactly 63, but is actually 63.54.

To distinguish isotopes, chemists make use of mass numbers. A copper atom with 29 protons and 34 neutrons has a mass number of 29 plus 34, or 63, and can therefore be referred to as "copper-63," while one with 29 protons and 36 neutrons would be "copper-65." In written form, chemical symbols plus superscripts are used, as Cu^{63} and Cu^{65}.

By this system, only the total number of protons plus neutrons are given. Chemists shrug this off. They know the atomic number of each element by heart (or they can look it up when no one's watching them), and that gives them the number of protons in the nucleus. By subtracting the atomic number from the mass number, they get the number of neutrons.

But for our purposes, I am going to write isotopes with proton and neutron numbers both clearly stated, thus: copper-29/34 and copper-29/36. If I want to refer to both of them, I will write: copper-29/34,36. Fair enough?

With this background, we can now look at the isotopes more closely. For instance, we can divide them into three varieties. First, there are the radioactive ones that break down so rapidly (lasting no longer than a few million years at most) that any which exist now have arisen in the comparatively near past as a result of some nuclear reaction, either in nature or in the laboratory. I will call these the "unstable" isotopes. Although over a thousand of these are known, each one exists in such fantastically small traces (if at all) that they make themselves known only to the nuclear physicist and his instruments.

Secondly, there are isotopes which are radioactive but which break down so slowly (in hundreds of millions of years at the very least) that those which exist today have existed at least since the original formation of the earth. Each of them, despite its continuous breakdown, exists in nature in quantities that would make it detectable by old-fashioned, nineteenth-century chemical methods. I will call these the "semi-stable" isotopes.

Finally there are the isotopes which are not at all radioactive or are so feebly radioactive that even our most sensitive instruments cannot detect it. These are the "stable" isotopes.

In this chapter I shall concern myself only with the semi-stable and stable isotopes.

No less than 22 of the 105 elements[2] known today possess only unstable isotopes and therefore exist in nature either in insignificant traces
or not at all. These are listed in Table 6. Notice that all but two of
these elements exist at the very end of the known list of elements, with
atomic numbers running from 84 to 105. The only elements not on the
list, within that stretch, are elements number 90 (thorium) and 92
(uranium), both of which, you will note, have an even atomic number. On the other hand, there are two elements in the list with atomic
numbers below that range, elements number 43 (technetium) and 61
(promethium), both with odd atomic numbers.

Table 6

ELEMENTS WITHOUT STABLE OR SEMI-STABLE ISOTOPES

ELEMENT	ATOMIC NUMBER	ELEMENT	ATOMIC NUMBER
Technetium	43	Americium	95
Promethium	61	Curium	96
Polonium	84	Berkelium	97
Astatine	85	Californium	98
Radon	86	Einsteinium	99
Francium	87	Fermium	100
Radium	88	Mendelevium	101
Actinium	89	Nobelium	102
Protactinium	91	Lawrencium	103
Neptunium	93	Rutherfordium	104
Plutonium	94	Hahnium	105

This means that there are exactly 83 elements which possess at least
one stable or semi-stable isotope and which therefore occur in reasonable quantities on earth. (There is no stable or semi-stable isotope that
does not occur in nature in reasonable quantities.) Some of these elements possess only one such isotope, some two, some three, and some
more than three.

Now, it is an odd thing that although every chemistry textbook I have
ever seen always lists the elements, no book I have ever seen lists the
isotopes in any systematic way.

[2] *I have adjusted these figures, and Table 6, to take into account the two newly
discovered elements 104 and 105.*

For instance, I have never seen *anywhere* a complete list of all those elements possessing but a single stable or semi-stable isotope. I have prepared such a list (Table 7).

Table 7

ELEMENTS WITH ONE STABLE OR SEMI-STABLE ISOTOPE

ELEMENT	PROTON/ NEUTRON	ELEMENT	PROTON/ NEUTRON
Beryllium	4/5	Rhodium	45/58
Fluorine	9/10	Iodine	53/74
Sodium	11/12	Cesium	55/78
Aluminum	13/14	Praseodymium	59/82
Phosphorus	15/16	Terbium	65/94
Scandium	21/24	Holmium	67/98
Manganese	25/30	Thulium	69/100
Cobalt	27/32	Gold	79/118
Arsenic	33/42	Bismuth	83/126
Yttrium	39/50	Thorium	90/142*
Niobium	41/52		

* semi-stable

There are twenty-one elements with one stable or semi-stable isotope apiece, and you will notice that in every case but two (beryllium and thorium, first and last in the list) the solo isotopes have an odd number of protons in the nucleus and an even number of neutrons. These are the "odd/even" isotopes.

Let's next list the elements that possess two stable or semi-stable isotopes (Table 8). This list includes twenty-three elements, of which twenty possess odd numbers of protons.

If you look at the first three tables, you will see that of the 53 known elements with odd atomic numbers, 13 possess no stable or semi-stable isotopes, 19 possess just one stable or semi-stable isotope and 20 possess just two. The total comes to 52.

There is one and only one element of odd atomic number left unaccounted for; and if you follow down the lists, the missing element turns out to be number 19, which is potassium. Potassium has three stable or semi-stable isotopes, and I'll list it here without giving it the dignity of a table all to itself: Potassium-19/20,21*,22. (The asterisk after the neutron number denotes a semi-stable isotope.)

Table 8

ELEMENTS WITH TWO STABLE OR
SEMI-STABLE ISOTOPES

ELEMENT	PROTON/ NEUTRONS	ELEMENT	PROTON/ NEUTRONS
Hydrogen	1/0,1	Silver	47/60,62
Helium	2/1,2	Indium	49/64,66*
Lithium	3/3,4	Antimony	51/70,72
Boron	5/5,6	Lanthanum	57/61*,62
Carbon	6/6,7	Europium	63/88,90
Nitrogen	7/7,8	Lutetium	71/104,105*
Chlorine	17/18,20	Tantalum	73/107*,108
Vanadium	23/27*,28	Rhenium	75/110,112*
Copper	29/34,36	Iridium	77/114,116
Gallium	31/38,40	Thallium	81/122,124
Bromine	35/44,46	Uranium	92/143*,146*
Rubidium	37/48,50*		

* semi-stable

Of these, the semi-stable potassium-19/21* (the lightest of all the semi-stable isotopes) makes up only one atom in every ten thousand of potassium, so that this element just *barely* has more than two isotopes.

The 53 elements with odd atomic numbers contain, all told, 62 different stable or semi-stable isotopes. Of these, 53 contain an even number of neutrons, so that there are 53 odd/even stable or semi-stable isotopes in existence. These can be broken up into 50 stable and 3 semi-stable (rubidium-37/50*, indium-49/66*, and rhenium-75/112*).

There are only nine stable or semi-stable isotopes of the atoms with odd atomic number that possess an odd number of neutrons as well. Table 9 contains a complete list of all the "odd/odd" isotopes, which are stable or semi-stable, in existence.

As you see, of these 9, fully 5 are semi-stable. This means that only 4 completely stable odd/odd isotopes exist in the universe. Of these, the odd/odd hydrogen-1/1 is outnumbered by the odd/even hydrogen-1/0 (I am calling zero an even number, if you don't mind) ten thousand to one. The odd/odd lithium-3/3 is outnumbered by the odd/even lithium-3/4 by thirteen to one, and the odd/odd boron-5/5 is outnumbered by the odd/even boron-5/6 by four to one. So three of the four stable odd/odd isotopes form minorities within their own elements.

Table 9

THE STABLE OR SEMI-STABLE
ODD/ODD ISOTOPES

ELEMENT	PROTON/NEUTRON
Hydrogen	1/1
Lithium	3/3
Boron	5/5
Nitrogen	7/7
Potassium	19/21*
Vanadium	23/27*
Lanthanum	57/61*
Lutetium	71/105*
Tantalum	73/107*

* *semi-stable*

This leaves nitrogen-7/7, an odd/odd isotope which is not only completely stable but which makes up 99.635 per cent of all nitrogen atoms. It is, in this respect, the oddest of all the odd/odds.

What about the elements of even atomic number?

There the situation is reversed. Only nine of the elements of even atomic number have no stable or semi-stable isotopes, and all of these are in the region beyond atomic number 83 where no fully stable and almost no semi-stable isotopes exist. What's more, the three semi-stable isotopes that do exist in that region all belong to elements of even atomic number.

There are two other elements of even atomic number with but a single stable or semi-stable isotope and three with but two stable isotopes. You can pick all these up in the tables already presented.

This leaves 39 of the 52 elements of even atomic number, all possessing more than two stable isotopes. One of them, tin, possesses no less than ten stable isotopes. I will not tabulate these elements in detail.

Instead, I will point out that there are two varieties of isotopes where elements of even atomic number are involved. There are isotopes with odd numbers of neutrons ("even/odd") and those with even numbers ("even/even").

The data on stable and semi-stable isotopes is summarized in Table 10.

In sheer numbers of isotopes, the even/even group is preponderant, making up 60 per cent of the total. The preponderance is even greater in mass.

Table 10

VARIETIES OF ISOTOPES

	STABLE	SEMI-STABLE	TOTAL
Even/Even	164	3	167
Even/Odd	55	2	57
Odd/Even	50	3	53
Odd/Odd	4	5	9
Total	273	13	
		Grand Total	286

Among the 43 elements of even atomic number that possess stable or semi-stable isotopes, only one lacks an even/even isotope. That is beryllium, with but one stable or semi-stable isotope, beryllium-4/5, which is even/odd.

Of the 42 others, there is not one case in which the even/even isotopes do not make up most of the atoms. The even/odd isotope which is most common within its own element is platinum-78/117, which makes up one-third of all platinum atoms. Where an element of even atomic number has more than one even/odd isotope (tin has three), all of them together sometimes do even better. The record is the case of xenon-54/75 and xenon-54/77, which together make up almost 48 per cent of all xenon atoms. In no case do the even/odd isotopes top the 50 per cent mark, except in the case of beryllium, of course.

What's more, the even/odd isotopes do best just in those elements which are least common. Platinum and xenon are among the rarest of all the elements with stable or semi-stable isotopes. It is precisely in the most common elements that the even/even isotopes are most predominant.

This shows up when we consider the structure of the Earth's crust. I once worked out its composition in isotope varieties and this is the result:

even/even — 85.63 per cent
odd/even — 13.11 per cent
even/odd — 1.25 per cent
odd/odd — 0.01 per cent

Almost 87 per cent of the Earth's crust is made up of the elements with even atomic numbers. And if the entire Earth is considered, the situation is even more extreme. Six elements make up 98 per cent of the globe, these being iron, oxygen, magnesium, silicon, sulfur and nickel. Every one of these is an element of even atomic number. I estimate that the globe we live on is 96 per cent even/even.

Which is a shame, in a way. As a long-time science-fiction enthusiast and practicing nonconformist, I have always had a sneaking sympathy for the odd/odd.

PART III

ORGANIC CHEMISTRY

8

YOU, TOO, CAN SPEAK GAELIC

It is difficult to prove to the man in the street that one is a chemist. At least, when one is a chemist after my fashion (strictly armchair).

Faced with a miscellaneous stain on a garment of unknown composition, I am helpless. I say "Have you tried a dry cleaner?" with a rising inflection that disillusions everyone within earshot at once. I cannot look at a paste of dubious composition and tell what it is good for just by smelling it; and I haven't the foggiest notion what a drug, identified only by trade name, may have in it.

It is not long, in short, before the eyebrows move upward, the wise smiles shoot from lip to lip, and the hoarse whispers begin: "Some chemist! Wonder what barber college *he* went to?"

There is nothing to do but wait. Sooner or later, on some breakfast-cereal box, on some pill dispenser, on some bottle of lotion, there will appear an eighteen-syllable name of a chemical. Then, making sure I have a moment of silence, I will say carelessly, "Ah, yes," and rattle it off like a machine gun, reducing everyone for miles around to stunned amazement.

Because, you see, no matter how inept I may be at the practical aspects of chemistry, I speak the language fluently.

But, alas, I have a confession to make. It isn't hard to speak chemistry. It just looks hard because organic chemistry (that branch of chemistry with the richest supply of nutcracker names) was virtually a German monopoly in the nineteenth century. The Germans, for some reason known only to themselves, push words together and eradicate all traces of any seam between them. What we would express as a phrase, they treat as one interminable word. They did this to the names of their organic compounds and in English those names were slavishly adopted with minimum change.

It is for that reason, then, that you can come up to a perfectly respectable compound which, to all appearances, is just lying there, harming no one, and find that it has a name like para-dimethylaminobenzaldehyde. (And that is rather short, as such names go.)

To the average person, used to words of a respectable size, this conglomeration of letters is offensive and irritating, but actually, if you tackle it from the front and work your way slowly toward the back, it isn't bad. Pronounce it this way: PA-ruh-dy-METH-il-a-MEE-noh-ben-ZAL-duh-hide. If you accent the capitalized syllables, you will discover that after a while you can say it rapidly and without trouble and you can impress your friends no end.

What's more, now that you can say the word, you will appreciate something that once happened to me. I was introduced to this particular compound some years ago, because when dissolved in hydrochloric acid, it is used to test for the presence of a compound called glucosamine and this was something I earnestly yearned to do at the time.

So I went to the reagent shelf and said to someone, "Do we have any para-dimethylaminobenzaldehyde?"

And he said, "What you mean is PA-ruh-dy-METH-il-a-MEE-noh-ben-ZAL-duh-hide," and he sang it to the tune of the "Irish Washerwoman."

If you don't know the tune of the "Irish Washerwoman," all I can say is that it is an Irish jig; in fact, it is *the* Irish jig; if you heard it, you would know it. I venture to say that if you know only one Irish jig, or if you try to make up an Irish jig, that's the one.

It goes: DUM-dee-dee-DUM-dee-dee-DUM-dee-dee-DUM-dee-dee, and so on almost indefinitely.

For a moment I was flabbergasted and then, realizing the enormity

of having someone dare be whimsical at my expense, I said. "Of course!
It's dactylic tetrameter."

"What?" he said.

I explained. A dactyl is a set of three syllables of which the first is ac-
cented and the next two are not, and a line of verse is dactylic tetrameter
when four such sets of syllables occur in it. Anything in dactylic feet can
be sung to the tune of the "Irish Washerwoman." You can sing most of
Longfellow's "Evangeline" to it, for instance, and I promptly gave the
fellow a sample:

"THIS is the FO-rest pri-ME-val. The MUR-muring PINES and the
HEM-locks—" and so on and so on.

He was walking away from me by then, but I followed him at a half
run. In fact, I went on, anything in iambic feet can be sung to the
tune of Dvorak's "Humoresque." (You know the one—dee-DUM-dee-
DUM-dee-DUM-dee-DUM-dee-DUM—and so on forever.)

For instance, I said, you could sing Portia's speech to the "Humoresque" like this: "The QUALiTY of MERcy IS not STRAINED it
DROPpeth AS the GENtle RAIN from HEAV'N uPON the PLACE
beNEATH."

He got away from me by then and didn't show up at work again for
days, and served him right.

However, I didn't get off scot free myself. Don't think it. I was
haunted for weeks by those drumming dactylic feet. PA-ruh-dy-METH-
il-a-MEE-noh-ben-ZAL-duh-hide-PA-ruh-dy-METH-il-a-MEE-noh—
went my brain over and over. It scrambled my thoughts, interfered
with my sleep, and reduced me to mumbling semimadness, for I would
go about muttering it savagely under my breath to the alarm of all in-
nocent bystanders.

Finally, the whole thing was exorcized and it came about in this
fashion. I was standing at the desk of a receptionist waiting for a chance
to give her my name in order that I might get in to see somebody. She
was a very pretty Irish receptionist and so I was in no hurry because the
individual I was to see was very masculine and I preferred the receptionist. So I waited patiently and smiled at her; and then her patent
Irishness stirred that drumbeat memory in my mind, so that I sang in a
soft voice (without even realizing what I was doing) PA-ruh-dy-METH-
il-a-MEE-noh-ben-ZAL-duh-hide . . . through several rapid choruses.

And the receptionist clapped her hands together in delight and cried
out, "*Oh, my, you know it in the original Gaelic!*"

JUSTUS VON LIEBIG

Liebig was born in Darmstadt, Hesse (then a small, independent German nation), on May 12, 1803. In 1818 he was apprenticed to an apothecary, but he did not rest until he could go to a university for formal instruction in chemistry. Arrested for political activity on the side of liberalism, Liebig had to leave Germany and managed to make his way to Paris.

Liebig worked on a series of compounds called "fulminates" while another German chemist, Friedrich Wöhler, was working on another called "cyanates." Each prepared papers that were published in the same journal and the editor noticed that although the two sets of

What could I do? I smiled modestly and had her announce me as Isaac O'Asimov.

From that day to this I haven't sung it once except in telling this story. It was gone, for after all, folks, in my heart I know I don't know one word of Gaelic.

But what are these syllables that sound so Gaelic? Let's trace them to their lair, one by one, and make sense of them, if we can. Perhaps you will then find that you, too, can speak Gaelic.

Let's begin with a tree of southeast Asia, one that is chiefly found in Sumatra and Java. It exudes a reddish-brown resin that, on being burned yields a pleasant odor. Arab traders had penetrated the Indian Ocean and its various shores during medieval times and had brought back this resin, which they called "Javanese incense." Of course, they called it that in Arabic, so that the phrase came out *"luban javi."*

When the Europeans picked up the substance from Arabic traders, the Arabic name was just a collection of nonsense syllables to them. The

compounds had different properties, they had the same elementary composition. They were "isomers," the first to be discovered. Liebig and Wöhler became close life-long friends in consequence.

Liebig also worked out a scheme for analyzing organic compounds carefully, determining the proportion of each element present with great accuracy. The existence of isomers made this necessary, since the equality of proportion demonstrated their existence.

In 1824 he began to teach at the university in the small Hessian city of Giessen and there he proved himself one of the great chemistry teachers of all time. He established a laboratory for general student use (something new in scientific education) and was the intellectual father and grandfather of most of the chemists since his time.

He exerted his analytical ability on tissue fluids and was the first to point out that carbohydrate and fat were the fuel of the body. He was the first to experiment with fertilization through the addition of chemicals to the soil in the place of natural products such as manures. The use of chemical fertilizers has not only greatly multiplied the food supply of those nations making use of scientific agriculture, but has also helped reduce epidemics through the elimination of the ubiquitous manure pile.

He died in Munich, Bavaria, on April 18, 1873.

first syllable *lu* sounded as though it might be the definite article (*lo* is one of the words for *the* in Italian; *le* and *la* are *the* in French and so on). Consequently, the European traders thought of the substance as "the banjavi" or simply as "banjavi."

That made no sense, either, and it got twisted in a number of ways; to "benjamin," for instance (because that, at least, was a familiar word), to "benjoin," and then finally, about 1650, to "benzoin." In English, the resin is now called "gum benzoin."

About 1608 an acid substance was isolated from the resin and that was eventually called "benzoic acid." Then, in 1834, a German chemist Eilhart Mitscherlich converted benzoic acid (which contains two oxygen atoms in its molecule) into a compound which contains no oxygen atoms at all, but only carbon and hydrogen atoms. He named the new compound "benzin," the first syllable signifying its ancestry.

Another German chemist Justus von Liebig objected to the suffix *-in*, which, he said, was used only for compounds that contained nitrogen atoms, which Mitscherlich's "benzin" did not. In this, Liebig was correct. However, he suggested the suffix *-ol*, signifying the German word for "oil," because the compound mixed with oils rather than with water. This was as bad as *-in*, however, for, as I shall shortly explain, the suffix *-ol* is used for other purposes by chemists. However, the name caught on in Germany, where the compound is still referred to as "benzol."

In 1845 still another German chemist (I told you organic chemistry was a German monopoly in the nineteenth century), August W. von Hofmann, suggested the name *benzene*, and this is the name properly used in most of the world, including the United States. I say properly, because the *-ene* ending is routinely used for many molecules containing hydrogen and carbon atoms only ("hydrocarbons") and therefore it is a good ending and a good name.

The molecule of benzene consists of six carbon atoms and six hydrogen atoms. The carbon atoms are arranged in a hexagon and to each of them is attached a single hydrogen atom. If we remember the actual structure we can content ourselves with stating that the formula of benzene is C_6H_6.

You will have noted, perhaps, that in the long and tortuous pathway from the island of Java to the molecule of benzene, the letters of the island have been completely lost. There is not a *j*, not an *a*, and not a *v*, in the word *benzene*.

Nevertheless, we've arrived somewhere. If you go back to the "Irish

Washerwoman" compound, para-dimethylaminobenzaldehyde, you will not fail to note the syllable *benz*. Now you know where it comes from.

Having got this far, let's start on a different track altogether.

Women, being what they are (three cheers), have for many centuries been shading their eyelashes and upper eyelids and eye corners in order to make said eyes look large, dark, mysterious, and enticing. In ancient times they used for this purpose some dark pigment (an antimony compound, often) which was ground up into a fine powder. It had to be a *very* fine powder, of course, because lumpy shading would look awful.

The Arabs, with an admirable directness, referred to this cosmetic powder as "the finely divided powder." Only, once again, they used Arabic and it came out "*al-kuhl*," where the *h* is pronounced in some sort of guttural way I can't imitate, and where *al* is the Arabic word for *the*.

The Arabs were the great alchemists of the early Middle Ages and when the Europeans took up alchemy in the late Middle Ages, they adopted many Arabic terms. The Arabs had begun to use *al-kuhl* as a name for any finely divided powder, without reference to cosmetic needs, and so did the Europeans. But they pronounced the word, and spelled it, in various ways that were climaxed with "alcohol."

As it happened, alchemists were never really at ease with gases or vapors. They didn't know what to make of them. They felt, somehow, that the vapors were not quite material in the same sense that liquids or solids were, and so they referred to the vapors as "spirits." They were particularly impressed with substances that gave off "spirits" even at ordinary temperatures (and not just when heated), and of these, the most important in medieval times was wine. So alchemists would speak of "spirits of wine" for the volatile component of wine (and we ourselves may speak of alcoholic beverages as "spirits," though we will also speak of "spirits of turpentine").

Then, too, when a liquid vaporizes it seems to powder away to nothing, so spirits also received the name *alcohol* and the alchemists would speak of "alcohol of wine." By the seventeenth century the word *alcohol* all by itself stood for the vapors given off by wine.

In the early nineteenth century the molecular structure of these vapors was determined. The molecule turned out to consist of two carbon atoms and an oxygen atom in a straight line. Three hydrogen atoms are attached to the first carbon, two hydrogen atoms to the second, and a single hydrogen atom to the oxygen. The formula can therefore be written as CH_3CH_2OH.

The hydrogen-oxygen group (-OH) is referred to in abbreviated form as a "hydroxyl group." Chemists began to discover numerous compounds in which a hydroxyl group is attached to a carbon atom, as it is in the alcohol of wine. All these compounds came to be referred to generally as alcohols, and each was given a special name of its own.

For instance, the alcohol of wine contains a group of two carbon atoms to which a total of five hydrogen atoms are attached. This same combination was discovered in a compound first isolated in 1540. This compound is even more easily vaporized than alcohol and the liquid disappears so quickly that it seems to be overwhelmingly eager to rise to its home in the high heavens. Aristotle had referred to the material making up the high heavens as "aether," so in 1730 this easily vaporized material received the name *spiritus aethereus*, or, in English, "ethereal spirits." This was eventually shortened to "ether."

The two-carbon-five-hydrogen group in either (there were two of these in each ether molecule) was naturally called the "ethyl group," and since the alcohol of wine contained this group, it came to be called "ethyl alcohol" about 1850.

It came to pass, then, that chemists found it sufficient to give the name of a compound the suffix -*ol* to indicate that it was an alcohol, and possessed a hydroxyl group. That is the reason for the objection to *benzol* as a name for the compound C_6H_6. Benzene contains no hydroxyl group and is not an alcohol and should be called "benzene" and not "benzol." You hear?

It is possible to remove two hydrogen atoms from an alcohol, taking away the single hydrogen that is attached to the oxygen, and one of the hydrogens attached to the adjoining carbon. Instead of the molecule CH_3CH_2OH, you would have the molecule CH_3CHO.

Liebig (the man who had suggested the naughty word *benzol*) accomplished this task in 1835 and was the first actually to isolate CH_3CHO. Since the removal of hydrogen atoms is, naturally, a "dehydrogenation," what Liebig had was a dehydrogenated alcohol, and that's what he called it. Since he used Latin, however, the phrase was *alcohol dehydrogenatus*.

That is a rather long name for a simple compound, and chemists, being as human as the next fellow (honest!), have the tendency to shorten long names by leaving out syllables. Take the first syllable of

alcohol and the first two syllables of *dehydrogenatus,* run the result together, and you have *aldehyde.*

Thus, the combination of a carbon, hydrogen, and oxygen atom
(-CHO), which forms such a prominent portion of the molecule of
dehydrogenated alcohol, came to be called the "aldehye group," and
any compound containing it came to be called an "aldehyde."

For instance, if we return to benzene, C_6H_6, and imagine one of
its hydrogen atoms removed, and in its place a -CHO group inserted,
we would have C_6H_5CHO and that compound would be "benzene-
aldehyde" or, to use the shortened form that is universally employed,
"benzaldehyde."

Now let's move back in time again to the ancient Egyptians. The
patron god of the Egyptian city of Thebes on the Upper Nile was named
Amen or Amun. When Thebes gained hegemony over Egypt, as it did
during the eighteenth and nineteenth dynasties, the time of Egypt's
greatest military power, Amen naturally gained hegemony over the
Egyptian gods. He rated many temples, including one on an oasis in
the North African desert, well to the west of the main center of Egyptian culture. This one was well known to the Greeks and, later, to the
Romans, who spelled the name of the god "Ammon."

Any desert area has a problem when it comes to finding fuel. One
available fuel in North Africa is camel dung. The soot of the burning
camel dung, which settled out on the walls and ceiling of the temple,
contained white, salt-like crystals, which the Romans then called "sal
ammoniac," meaning "salt of Ammon." (The expression "sal ammoniac" is still good pharmacist's jargon, but chemists call the substance
"ammonium chloride" now.)

In 1774 an English chemist Joseph Priestley discovered that heating
sal ammoniac produced a vapor with a pungent odor, and in 1782 the
Swedish chemist Torbern Olof Bergman suggested the name *ammonia*
for this vapor. Three years later a French chemist, Claude Louis
Berthollet, worked out the structure of the ammonia molecule. It consisted of a nitrogen atom to which three hydrogen atoms were attached,
so that we can write it NH_3.

As time went on, chemists who were studying organic compounds
(that is, compounds that contained carbon atoms) found that it often
happened that a combination made up of a nitrogen atom and two
hydrogen atoms (-NH$_2$) was attached to one of the carbon atoms in
the organic molecule. The resemblance of this combination to the am-

monia molecule was clear, and by 1860 the $-NH_2$ group was being called an "amine group" to emphasize the similarity.

Well, then, if we go back to our benzaldehyde, C_6H_5CHO, and imagine a second hydrogen atom removed from the original benzene and in its place an amine group inserted, we would have $C_6H_4(CHO)$ (NH_2) and that would be "aminobenzaldehyde."

Earlier I talked about the alcohol of wine, CH_3CH_2OH, and said it was "ethyl alcohol." It can also be called (and frequently is) "grain alcohol" because it is obtained from the fermentation of grain. But, as I hinted, it is not the only alcohol; far from it. As far back as 1661, the English chemist Robert Boyle found that if he heated wood in the absence of air, he obtained vapors, some of which condensed into clear liquid.

In this liquid he detected a substance rather similar to ordinary alcohol, but not quite the same. (It is more easily evaporated than ordinary alcohol, and it is considerably more poisonous, to mention two quick differences.) This new alcohol was called "wood alcohol."

However, for a name really to sound properly authoritative in science, what is really wanted is something in Greek or Latin. The Greek word for wine is *methy* and the Greek work for wood is *hylē*. To get "wine from wood" (i.e., "wood alcohol"), stick the two Greek words together and you have *methyl*. The first to do this was the Swedish chemist Jöns Jakob Berzelius, about 1835, and ever since then wood alcohol has been "methyl alcohol" to chemists.

The formula for methyl alcohol was worked out in 1834 by a French chemist named Jean Baptiste André Dumas (no relation to the novelist, as far as I know). It turned out to be simpler than that of ethyl alcohol and to contain but one carbon atom. The formula is written CH_3OH. For this reason, a grouping of one carbon atom and three hydrogen atoms ($-CH_3$) came to be referred to as a "methyl group."

The French chemist Charles Adolphe Wurtz (he was born in Alsace, which accounts for his Germanic name) discovered in 1849 that one of the two hydrogen atoms of the amine group could be replaced by a methyl group, so that the end product looked like this: $-NHCH_3$. This would naturally be a "methylamine group." If both hydrogen atoms were replaced by methyl groups, the formula would be $-N(CH_3)_2$ and we would have a "dimethylamine group." (The prefix *di-* is from the Greek *dis*, meaning "twice." The methyl group is added to the amine group twice, in other words.)

Now we can go back to our aminobenzaldehyde, $C_6H_4(CHO)$-(NH_2). If, instead of an amine group, we had used a dimethylamine group, the formula would be $C_6H_4(CHO)(N(CH_3)_2)$ and the name would be "dimethylaminobenzaldehyde."

Let's think about benzene once again. Its molecule is a hexagon made up of six carbon atoms, each with a hydrogen atom attached. We have substituted an aldehyde group for one of the hydrogen atoms and a dimethylamine group for another, to form dimethylaminobenzaldehyde, but *which* two hydrogen atoms have we substituted?

In a perfectly symmetrical hexagon, such as that which is the molecule of benzene, there are only three ways in which you can choose two hydrogen atoms. You can take the hydrogen atoms of two adjoining carbon atoms; or you can take the hydrogens of two carbon atoms so selected that one untouched carbon-hydrogen combination lies between; or you can take them so that two untouched carbon-hydrogen combinations lie between.

If you number the carbon atoms of the hexagon in order, one through six, then the three possible combinations involve carbons 1,2; 1,3; and 1,4 respectively. If you draw a diagram for yourself (it is simple enough), you will see that no other combinations are possible. All the different combinations of two carbon atoms in the hexagon boil down to one or another of these three cases.

Chemists have evolved a special name for each combination. The 1,2 combination is *ortho* from a Greek word meaning "straight" or "correct," perhaps because it is the simplest in appearance, and what seems simple, seems correct.

The prefix *meta*- comes from a Greek word meaning "in the midst of," but it also has a secondary meaning, "next after." That makes it suitable for the 1,3 combination. You substitute the first carbon, leave the next untouched, and substitute the one "next after."

The prefix *para*- is from a Greek word meaning "beside" or "side by side." If you mark the 1,4 angles on a hexagon and turn it so that the 1 is at the extreme left, then the 4 will be at the extreme right. The two are indeed "side by side" and so *para*- is used for the 1,4 combination.

Now we know where we are. When we say "para-dimethylaminobenzaldehyde," we mean that the dimethylamine group and the aldehyde group are in the 1,4 relationship to each other. They are at opposite ends of the benzene ring and we can write the formula $CHOC_6H_4N(CH_3)_2$.

See?

Now that you know Gaelic, what do you suppose the following are?

(1) alpha-dee-glucosido-beta-dee-fructofuranoside

(2) two, three-dihydro-three-oxobenzisosulfonazole

(3) delta-four-pregnene-seventeen-alpha, twenty-one, diol-three, eleven, twenty-trione

(4) three-(four-amino-two-methylpyrimidyl-five-methyl)-four-methyl-five-beta-hydroxyethylthiazolium chloride hydrochloride

Just in case your Gaelic is still a little rusty, I will give you the answers. There are:

                    (1) table sugar

                    (2) saccharin

                    (3) cortisone

                    (4) vitamin B_1

Isn't it simple?

PART IV

BIOCHEMISTRY

9

THE HASTE-MAKERS

When I first began writing about science for the general public—far back in medieval times—I coined a neat phrase about the activity of a "light-fingered magical catalyst."

My editor stiffened as he came across that phrase, but not with admiration (as had been my modestly confident expectation). He turned on me severely and said, "Nothing in science is magical. It may be puzzling, mysterious, inexplicable—but it is *never magical*."

It pained me, as you can well imagine, to have to learn a lesson from an *editor*, of all people, but the lesson seemed too good to miss and, with many a wry grimace, I learned it.

That left me, however, with the problem of describing the workings of a catalyst, without calling upon magical power for an explanation.

Thus, one of the first experiments conducted by any beginner in a high school chemistry laboratory is to prepare oxygen by heating potassium chlorate. If it were only potassium chlorate he were heating, oxygen would be evolved but slowly and only at comparatively high temperatures. So he is instructed to add some manganese dioxide first.

When he heats the mixture, oxygen comes off rapidly at comparatively low temperatures.

What does the manganese dioxide do? It contributes no oxygen. At the conclusion of the reaction it is all still there, unchanged. Its mere presence seems sufficient to hasten the evolution of oxygen. It is a hastemaker or, more properly, a catalyst.

And how can one explain influence by mere presence? Is it a kind of molecular action at a distance, an extrasensory perception on the part of potassium chlorate that the influential aura of manganese dioxide is present? Is it telekinesis, a para-natural action at a distance on the part of the manganese dioxide? Is it, in short, magic?

Well, let's see . . .

To begin at the beginning, as I almost invariably do, the first and most famous catalyst in scientific history never existed.

The alchemists of old sought methods for turning base metals into gold. They failed, and so it seemed to them that some essential ingredient was missing in their recipes. The more imaginative among them conceived of a substance which, if added to the mixture they were heating (or whatever) would bring about the production of gold. A small quantity would suffice to produce a great deal of gold and it could be recovered and used again, no doubt.

No one had ever seen this substance but it was described, for some reason, as a dry, earthy material. The ancient alchemists therefore called it *xerion*, from a Greek word meaning "dry."

In the eighth century the Arabs took over alchemy and called this gold-making catalyst "the xerion" or, in Arabic, *al-iksir*. When West Europeans finally learned Arabic alchemy in the thirteenth century, *al-iksir* became "elixir."

As a further tribute to its supposed dry, earthy properties, it was commonly called, in Europe, "the philosopher's stone." (Remember that as late as 1800, a "natural philosopher" was what we would now call a "scientist.")

The amazing elixir was bound to have other marvelous properties as well, and the notion arose that it was a cure for all diseases and might very well confer immortality. Hence, alchemists began to speak of "the elixir of life."

For centuries, the philosopher's stone and/or the elixir of life was searched for but not found. Then, when finally a catalyst was found, it brought about the formation not of lovely, shiny gold, but messy, dangerous sulfuric acid.[1] Wouldn't you know?

Before 1740, sulfuric acid was hard to prepare. In theory, it was easy. You burn sulfur, combining it with oxygen to form sulfur dioxide (SO_2). You burn sulfur dioxide further to make sulfur trioxide (SO_3). You dissolve sulfur trioxide in water to make sulfuric acid (H_2SO_4). The trick, though, was to make sulfur dioxide combine with oxygen. That could only be done slowly and with difficulty.

In the 1740s, however, an English sulfuric acid manufacturer named Joshua Ward must have reasoned that saltpeter (potassium nitrate), though nonflammable itself, caused carbon and sulfur to burn with great avidity. (In fact, carbon plus sulfur plus saltpeter is gunpowder.) Consequently, he added saltpeter to his burning sulfur and found that he now obtained sulfur trioxide without much trouble and could make sulfuric acid easily and cheaply.

The most wonderful thing about the process was that, at the end, the saltpeter was still present, unchanged. It could be used over and over again. Ward patented the process and the price of sulfuric acid dropped to 5 per cent of what it was before.

Magic? —Well, no.

In 1806, two French chemists, Charles Bernard Désormes and Nicholas Clément, advanced an explanation that contained a principle which is accepted to this day.

It seems, you see, that when sulfur and saltpeter burn together, sulfur dioxide combines with a portion of the saltpeter molecule to form a complex. The oxygen of the saltpeter portion of the complex transfers to the sulfur dioxide portion, which now breaks away as sulfur trioxide.

What's left (the saltpeter fragment minus oxygen) proceeds to pick up that missing oxygen, very readily, from the atmosphere. The saltpeter fragment, restored again, is ready to combine with an additional molecule of sulfur dioxide and pass along oxygen. It is the saltpeter's task simply to pass oxygen from air to sulfur dioxide as fast as it can. It is a middleman, and *of course* it remains unchanged at the end of the reaction.

In fact, the wonder is not that a catalyst hastens a reaction while remaining apparently unchanged, but that anyone should suspect even for a moment that anything "magical" is involved. If we were to come

[1] *That's all right, though. Sulfuric acid may not be as costly as gold, but it is —conservatively speaking—a trillion times as intrinsically useful.*

across the same phenomenon in the more ordinary affairs of life, we would certainly not make that mistake of assuming magic.

For instance, consider a half-finished brick wall and, five feet from it, a heap of bricks and some mortar. If that were all, then you would expect no change in the situation between 9 A.M. and 5 P.M. except that the mortar would dry out.

Suppose, however, that at 9 A.M. you observed one factor in addition—a man, in overalls, standing quietly between the wall and the heap of bricks with his hands empty. You observed matters again at 5 P.M. and the same man is standing there, his hands still empty. He has not changed. However, the brick wall is now completed and the heap of bricks is gone.

The man clearly fulfills the role of catalyst. A reaction has taken place as a result, apparently, of his mere presence and without any visible change or diminution in him.

Yet would we dream for a moment of saying "Magic!"? We would, instead, take it for granted that had we observed the man in detail all day, we would have caught him transferring the bricks from the heap to the wall one at a time. And what's not magic for the bricklayer is not magic for the saltpeter, either.

With the birth and progress of the nineteenth century, more examples of this sort of thing were discovered. In 1812, for instance, the Russian chemist Gottlieb Sigismund Kirchhoff . . .

And here I break off and begin a longish digression for no other reason than that I want to; relying, as I always do, on the infinite patience and good humor of the Gentle Readers.

It may strike you that in saying "the Russian chemist, Gottlieb Sigismund Kirchhoff" I have made a humorous error. Surely no one with a name like Gottlieb Sigismund Kirchhoff can be a Russian! It depends, however, on whether you mean a Russian in an ethnic or in a geographic sense.

To explain what I mean, let's go back to the beginning of the thirteenth century. At that time, the regions of Courland and Livonia, along the southeastern shores of the Baltic Sea (the modern Latvia and Estonia) were inhabited by virtually the last group of pagans in Europe. It was the time of the Crusades, and the Germans to the southeast felt it a pious duty to slaughter the poorly armed and dis organized pagans for the sake of their souls.

The crusading Germans were of the "Order of the Knights of the Sword" (better known by the shorter and more popular name of "Livonian Knights"). They were joined in 1237 by the Teutonic Knights, who had first established themselves in the Holy Land. By the end of the thirteenth century the Baltic shores had been conquered, with the German expeditionary forces in control.

The Teutonic Knights, as a political organization, did not maintain control for more than a couple of centuries. They were defeated by the Poles in the 1460s. The Swedes, under Gustavus Adolphus, took over in the 1620s, and in the 1720s the Russians, under Peter the Great, replaced the Swedes.

Nevertheless, however the political tides might shift and whatever flag flew and to whatever monarch the loyal inhabitants might drink toasts, the land itself continued to belong to the "Baltic barons" (or "Balts") who were the German-speaking descendants of the Teutonic Knights.

Peter the Great was an aggressive Westernizer who built a new capital, St. Petersburg[2] at the very edge of the Livonian area, and the Balts were a valued group of subjects indeed.

This remained true all through the eighteenth and nineteenth centuries when the Balts possessed an influence within the Russian Empire out of all proportion to their numbers. Their influence in Russian science was even more lopsided.

The trouble was that public education within Russia lagged far behind its status in western Europe. The Tsars saw no reason to encourage public education and make trouble for themselves. No doubt they felt instinctively that a corrupt and stupid government is only really safe with an uneducated populace.

This meant that even elite Russians who wanted a secular education had to go abroad, especially if they wanted a graduate education in science. Going abroad was not easy, either, for it meant learning a new language and new ways. What's more, the Russian Orthodox Church viewed all Westerners as heretics and little better than heathens. Contact with heathen ways (such as science) was at best dangerous and at worst damnation. Consequently, for a Russian to travel West for an education meant the overcoming of religious scruples as well.

The Balts, however, were German in culture and Lutheran in religion

2 *The city was named for his name-saint and not for himself. Whatever Tsar Peter was, a saint he was not.*

and had none of these inhibitions. They shared, with the Germans of Germany itself, in the heightening level of education—in particular, of scientific education—through the eighteenth and nineteenth centuries.

So it follows that among the great Russian scientists of the nineteenth century we not only have a man with a name like Gottlieb Sigismund Kirchhoff, but also others with names like Friedrich Konrad Beilstein, Karl Ernst von Baer, and Wilhelm Ostwald.

This is not to say that there weren't Russian scientists in this period with Russian names. Examples are Mikhail Vasilievich Lomonosov, Aleksandr Onufrievich Kovalevski, and Dmitri Ivanovich Mendeleev.

However, Russian officialdom actually preferred the Balts (who supported the Tsarist government under which they flourished) to the Russian intelligentsia itself (which frequently made trouble and had vague notions of reform).

In addition, the Germans were the nineteenth-century scientists *par excellence,* and to speak Russian with a German accent probably lent distinction to a scientist. (And before you sneer at this point of view, just think of the American stereotype of a rocket scientist. He has a thick German accent, *nicht wahr?* —And this despite the fact that the first rocketman, and the one whose experiments started the Germans on the proper track [Robert Goddard], spoke with a New England twang.)

JOHANN WOLFGANG DÖBEREINER

Döbereiner was born in Hof, Bavaria, December 15, 1780. He was the son of a coachman and received very little formal education. He read widely, however, attended any learned lecture he could get to, and somehow managed to display sufficient ability to attract the attention of a nobleman, who used his influence to obtain a position for him in 1810 as a professor of chemistry and physics at the University of Jena.

In the 1820s Döbereiner found that powdered platinum somehow influenced the reaction of hydrogen and oxygen. In fact, he invented an automatic lighter called "Döbereiner's lamp" that was based on this fact. It was an arrangement whereby a jet of hydrogen could be played, at will, upon powdered platinum, at which point the hydrogen would catch fire at once. The excitement over this didn't last long, for the platinum was too expensive to begin with and was quickly fouled by the impurities in the hydrogen and stopped working.

Nevertheless, Döbereiner's discovery was the precursor of growing knowledge of the manner in which certain substances, both inorganic and organic, could influence the course of reactions. This influence was called "catalysis."

Döbereiner also noticed some puzzling regularities in the list of elements which in his time had expanded in number to over fifty in no particular order. In 1829, he pointed out that the element bromine seemed just halfway in its properties between chlorine and iodine. It was also halfway between in atomic weight. He found other such "triads," such as calcium, strontium, and barium and sulfur, selenium, and tellurium.

This discovery was dismissed by most chemists as amusing, but unimportant. It proved, however, to be the precursor of the crucial working out of the periodic table by Mendeleev forty years later

Döbereiner did not live to see this, for he died in Jena on March 24, 1849.

So it happened that the Imperial Academy of Sciences of the Russian Empire (the most prestigious scientific organization in the land) was divided into a "German party" and a "Russian party," with the former dominant.

In 1880 there was a vacancy in the chair of chemical technology at the Academy, and two names were proposed. The German party proposed Beilstein, and the Russian party proposed Mendeleev. There was no comparison really. Beilstein spent years of his life preparing an encyclopedia of the properties and methods of preparation of many thousands of organic compounds which, with numerous supplements and additions, is still a chemical bible. This is a colossal monument to his thorough, hard-working competence—but it is no more. Mendeleev, who worked out the periodic table of the elements, was, on the other hand, a chemist of the first magnitude—an undoubted genius in the field.

Nevertheless, government officials threw their weight behind Beilstein, who was elected by a vote of ten to nine.

It is no wonder, then, that in recent years, when the Russians have finally won a respected place in the scientific sun, they tend to overdo things a bit. They've got a great deal of humiliation to make up for.

That ends the digression, so I'll start over—

As the nineteenth century wore on, more examples of haste-making were discovered. In 1812, for instance, the Russian chemist Gottlieb Sigismund Kirchhoff found that if he boiled starch in water to which a small amount of sulfuric acid had been added, the starch broke down to a simple form of sugar, one that is now called glucose. This would not happen in the absence of acid. When it did happen in the presence of acid, that acid was not consumed but was still present at the end.

Then, in 1816, the English chemist Humphry Davy found that certain organic vapors, such as those of alcohol, combined with oxygen more easily in the presence of metals such as platinum. Hydrogen combined more easily with oxygen in the presence of platinum also.

Fun and games with platinum started at once. In 1823 a German chemist, Johann Wolfgang Döbereiner, set up a hydrogen generator which, on turning an appropriate stopcock, would allow a jet of hydrogen to shoot out against a strip of platinum foil. The hydrogen promptly burst into flame and "Döbereiner's lamp" was therefore the first cigarette lighter. Unfortunately, impurities in the hydrogen gas quickly "poisoned" the expensive bit of platinum and rendered it useless.

In 1831 an English chemist, Peregrine Phillips, reasoned that if platinum could bring about the combination of hydrogen and of alcohol with oxygen, why should it not do the same for sulfur dioxide? Phillips found it would and patented the process. It was not for years afterward, however, that methods were discovered for delaying the poisoning of the metal, and it was only after that that a platinum catalyst could be profitably used in sulfuric acid manufacture to replace Ward's saltpeter.

In 1836 such phenomena were brought to the attention of the Swedish chemist Jöns Jakob Berzelius who, during the first half of the nineteenth century, was the uncrowned king of chemistry. It was he who suggested the words "catalyst" and "catalysis" from Greek words meaning "to break down" or "to decompose." Berzelius had in mind such examples of catalytic action as the decomposition of the large starch molecule into smaller sugar molecules by the action of acid.

But platinum introduced a new glamor to the concept of catalysis. For one thing, it was a rare and precious metal. For another, it enabled people to begin suspecting magic again.

Can platinum be expected to behave as a middleman as saltpeter does?

At first blush, the answer to that would seem to be in the negative. Of all substances, platinum is one of the most inert. It doesn't combine with oxygen or hydrogen under any normal circumstances. How, then, can it cause the two to combine?

If our metaphorical catalyst is a bricklayer, then platinum can only be a bricklayer tightly bound in a straitjacket.

Well, then, are we reduced to magic? To molecular action at a distance?

Chemists searched for something more prosaic. The suspicion grew during the nineteenth century that the inertness of platinum is, in one sense at least, an illusion. In the body of the metal, platinum atoms are attached to each other in all directions and are satisfied to remain so. In bulk, then, platinum will not react with oxygen or hydrogen (or most other chemicals, either).

On the surface of the platinum, however, atoms on the metal boundary and immediately adjacent to the air have no other platinum atoms, in the air-direction at least, to attach themselves to. Instead, then, they

JÖNS JAKOB BERZELIUS

Berzelius was born in Väversunda, Sweden on August 20, 1779. He lost his father, a schoolmaster, while very young and his mother when not very much older. During his mother's widowhood, she had married again, however, and the stepfather saw to the young man's education. He attended medical school, where he did poorly in medicine but very well in physics.

About 1807 he started a program of analyzing various inorganic compounds to determine the exact ratio of the weight of the different elements composing them. By running two thousand analyses over a period of ten years, he could present many examples of elementary proportions fixed in such a way as to back up the atomic theories of Dalton.

Berzelius was the second, after Dalton, to work up a table of atomic weights. Berzelius' table was, however, the first to be a really extensive one and a satisfactorily accurate one. In preparing his tables of elements, he avoided the tedium of writing the name of each attach themselves to whatever atoms or molecules they find handy—oxygen atoms, for instance. This forms a thin film over the surface, a film one molecule thick. It is completely invisible, of course, and all we see is a smooth, shiny, platinum surface, which seems completely non-reactive and inert.

As parts of a surface film, oxygen and hydrogen react more readily than they do when making up bulk gas. Suppose, then, that when a water molecule is formed by the combination of hydrogen and oxygen on the platinum surface, it is held more weakly than an oxygen molecule would be. The moment an oxygen molecule struck that portion of the surface it would replace the water molecule in the film. Now there would be the chance for the formation of another water molecule, and so on.

The platinum does act as a middleman after all, through its formation of the monomolecular gaseous film.

Furthermore, it is also easy to see how a platinum catalyst can be poisoned. Suppose there are molecules to which the platinum atoms will cling even more tightly than to oxygen. Such molecules will replace oxygen wherever it is found on the film and will not themselves be replaced by any gas in the atmosphere. They are on the platinum

element in full by working out the system of chemical symbols we still use today.

He also discovered a number of new elements: cerium, selenium, silicon, and thorium.

By 1830 Berzelius was the great chemical authority of the world. His textbook of chemistry, first published in 1803 and going through five editions before his death, was considered the last word. Between 1821 and 1849, he published a yearly review of chemical progress in which he editorialized on the work of others, and when he condemned a new suggestion or experiment, it was as good as dead. This was not altogether useful, for Berzelius grew conservative in his old age and held to his own ideas all the more fiercely when they were wrong.

At the age of fifty-six he finally married, taking to himself a good-looking twenty-four-year-old woman and being made a baron by the Swedish king as a wedding present. His last decade of life was spent happily and he died in Stockholm on August 7, 1848.

surface to stay, and any catalytic action involving hydrogen or oxygen is killed.

Since it takes very little substance to form a layer merely one molecule thick over any reasonable stretch of surface, a catalyst can be quickly poisoned by impurities that are present in the working mixture of gases, even when those impurities are present only in trace amounts.

If this is all so, then anything which increases the amount of surface in a given weight of metal will also increase the catalytic efficiency. Thus, powdered platinum, with a great deal of surface, is a much more effective catalytic agent than the same weight of bulk platinum. It is perfectly fair, therefore, to speak of "surface catalysis."

But what is there about a surface film that hastens the process of, let us say, hydrogen-oxygen combination? We still want to remove the suspicion of magic.

To do so, it helps to recognize what catalysts *can't* do.

For instance, in the 1870s, the American physicist Josiah Willard Gibbs painstakingly worked out the application of the laws of thermodynamics to chemical reactions. He showed that there is a quantity called "free energy" which always decreases in any chemical reaction that is spontaneous—that is, that proceeds without any input of energy.

Thus, once hydrogen and oxygen start reacting, they keep on reacting for as long as neither gas is completely used up, and as a result of the reaction water is formed. We explain this by saying that the free energy of the water is less than the free energy of the hydrogen-oxygen mixture. The reaction of hydrogen and oxygen to form water is analogous to sliding down an "energy slope."

But if that is so, why don't hydrogen and oxygen molecules combine with each other as soon as they are mixed? Why do they linger for indefinite periods at the top of the energy slope after being mixed, and react and slide downward only after being heated?

Apparently, before hydrogen and oxygen molecules (each composed of a pair of atoms) can react, one or the other must be pulled apart into individual atoms. That requires an energy input. It represents an upward energy slope, before the downward slope can be entered. It is an "energy hump," so to speak. The amount of energy that must be put into a reacting system to get it over that energy hump is called the "energy of activation," and the concept was first advanced in 1889 by the Swedish chemist Svante August Arrhenius.

When hydrogen and oxygen molecules are colliding at ordinary temperature, only the tiniest fraction happen to possess enough energy of motion to break up on collision. That tiniest fraction, which does break up and does react, then liberates enough energy, as it slides down the energy slope, to break up additional molecules. However, so little energy is produced at any one time that it is radiated away before it can do any good. The net result is that hydrogen and oxygen mixed at room temperature do not react.

If the temperature is raised, molecules move more rapidly and a larger proportion of them possess the necessary energy to break up on collision. (More, in other words, can slide over the energy hump.) More and more energy is released, and there comes a particular temperature when more energy is released than can be radiated away. The temperature is therefore further raised, which produces more energy, which raises the temperature still further—and hydrogen and oxygen proceed to react with an explosion.

In 1894 the Russian chemist Wilhelm Ostwald pointed out that a catalyst could not alter the free energy relationships. It cannot really make a reaction go, that would not go without it—though it can make a reaction go rapidly that in its absence would proceed with only imperceptible speed.

In other words, hydrogen and oxygen combine in the absence of platinum but at an imperceptible rate, and the platinum haste-maker accelerates that combination. For water to decompose to hydrogen and oxygen at room temperature (without the input of energy in the form of an electric current, for instance) is impossible, for that would mean spontaneously moving up an energy slope. Neither platinum nor any other catalyst could make a chemical reaction move up an energy slope. If we found one that did so, then *that* would be magic.[3]

But *how* does platinum hasten the reaction it does hasten? What does it do to the molecules in the film?

Ostwald's suggestion (accepted ever since) is that catalysts hasten reactions by lowering the energy of activation of the reaction—flattening out the hump. At any given temperature, then, more molecules can cross over the hump and slide downward, and the rate of the reaction increases, sometimes enormously.

For instance, the two oxygen atoms in an oxygen molecule hold together with a certain, rather strong, attachment, and it is not easy to

---

[3] *Or else we would have to modify the laws of thermodynamics.*

split them apart. Yet such splitting is necessary if a water molecule is to be formed.

When an oxygen atom is attached to a platinum atom and forms part of a surface film, however, the situation changes. Some of the bond-forming capabilities of the oxygen molecule are used up in forming the attachment to the platinum, and less is available for holding the two oxygen molecules together. The oxygen atom might be said to be "strained."

If a hydrogen atom happens to strike such an oxygen molecule, strained in the film, it is more likely to knock it apart into individual oxygen atoms (and react with one of them) than would be the case if it collided with an oxygen atom free in the body of a gas. The fact that the oxygen molecule is strained means that it is easier to break apart, and that the energy of activation for the hydrogen-oxygen combination has been lowered.

Or we can try a metaphor again. Imagine a brick resting on the upper reaches of a cement incline. The brick should, ideally, slide down the incline. To do so, however, it must overcome frictional forces which hold it in place against the pull of gravity. The frictional forces are here analogous to the forces holding the oxygen molecule together.

To overcome the frictional force one must give the brick an initial push (the energy of activation), and then it slides down.

Now, however, we will try a little "surface catalysis." We will coat the slide with wax. If we place the brick on top of such an incline, the merest touch will start it moving downward. It may move downward without any help from us at all.

In waxing the cement incline we haven't increased the force of gravity, or added energy to the system. We have merely decreased the frictional forces (that is, the energy hump), and bricks can be delivered down such a waxed incline much more easily and much more rapidly than down an unwaxed incline.

So you see that on inspection, the magical clouds of glory fade into the light of common day, and the wonderful word "catalyst" loses all its glamor. In fact, nothing is left to it but to serve as the foundation for virtually all of chemical industry and, in the form of enzymes, the foundation of all of life, too.

And, come to think of it, that ought to be glory enough for any reasonable catalyst.

10

LIFE'S BOTTLENECK

Villains on a cosmic scale are where you find them, and the imagination has found some majestic ones indeed, including exploding suns and invading Martians. Real life, in recent years, has found some actual villains that would have seemed most imaginary not too long ago, as, for instance, nuclear bombs and melting icecaps.

But there are always a few more, if you look long enough,—like sanitary plumbing and modern sewage disposal.

Well, let me explain.

To begin with, let's consider the ocean, the mother of all things living. Out of its substance, some billions of years ago, life formed, utilizing for the purpose the various types of atoms found in the ocean, though it had to juggle the proportions a bit.

For instance, the ocean is mainly water, and so is living tissue. The ocean is 97 per cent water by weight, while living things in the ocean are about 80 per cent water, generally speaking.

However, this is not quite a fair comparison. The water molecule is made up of two hydrogen atoms and an oxygen atom. In the ocean, water itself is the only substance, to speak of, which contains these atoms. In living matter, however, hydrogen and oxygen are contained in many of the constituent molecules other than water; and all this hydrogen and oxygen came from water originally. This "hydrogen-and-oxygen-elsewhere" should be counted with water, therefore.

To get a more proper panorama, let's consider the percentage by weight of each constituent type of atom. We can do this for the ocean, and for the copepod, a tiny crustacean which is one of the more common forms of the ocean's swarming animal life. The results are given in Table 11.

The column headed "Concentration Factor" in that table is the most important part of it. It represents the ratio of the percentage of a particular substance in living tissue to the percentage of the same substance in the environment.

Table 11

	PER CENT COMPOSITION OF OCEAN	PER CENT COMPOSITION OF COPEPOD	CONCENTRATION FACTOR
Oxygen	85.89	79.99	0.93
Hydrogen	10.82	10.21	0.94
Everything else	3.29	9.80	3.35

For instance, oxygen and hydrogen are found in smaller percentage in tissue than in ocean so the concentration factor for each is less than 1, as is shown in the table. To convert 100 pounds of ocean (containing 96.71 pounds of hydrogen and oxygen) into 100 pounds of copepod (containing 90.20 pounds of hydrogen and oxygen) 6.51 pounds of hydrogen and oxygen must be gotten rid of.

Whenever the concentration factor for any substance is less than 1, it means that that particular substance, potentially at least, can never be a limiting factor in the multiplication of living things. Life's problem will always be to get rid of it, rather than to collect it.

The situation is the reverse as far as "everything else" is concerned. Here 100 pounds of copepod contains 9.80 pounds of "everything else" while 100 pounds of ocean—out of which the copepod is formed—contains only 3.29 pounds. It takes 335 pounds of ocean to contain the 9.80 pounds of "everything else."

A concentration factor greater than 1 sets up the possibility of a bottleneck. Ideally, life could multiply in the ocean till the entire ocean had been converted into living tissue. After all, what is there to stop the endless and unlimited multiplication of life?

Well, suppose you begin with 335 pounds of ocean. By the time copepods have multiplied to a total weight of 100 pounds, they have incorporated all the "everything else" in the supply of ocean into their own bodies. There is still 235 pounds of ocean left, but it is pure water and cannot be converted into copepod.

The greater the concentration factor, the more quickly that limit would have been reached and the smaller the fraction of the total environment that can be converted into living tissue.

Of course, I have deliberately simplified the matter, to begin with, in order to make the point. Actually, the "everything else" is a conglomerate of a dozen or so elements, each of which is essential to life, and none of which can be dispensed with.

Each essential element is present in different amounts in the ocean; each is present in different amounts in living tissue. Each, therefore, has its own concentration factor. As soon as any one of them is completely used up, the possibility of the further expansion of life, generally, halts. One form of life can grow at the expense of another, but the total quantity of protoplasm can increase no further.

The essential element with the highest concentration factor is the one first used up and is, therefore, life's bottleneck.

Let's therefore compare the ocean and the copepod in finer detail, omitting the hydrogen and oxygen and just considering the "everything else." This is done in Table 12.

You can see that concentration factors do indeed vary widely from element to element. Only four elements have factors that are really extreme; that is, over a thousand. Of these four, the values for carbon and nitrogen are not really as extreme as they seem, however, since the ocean is not the only source of these elements. There is, for instance, some carbon dioxide in the air, and all of that is available to ocean life. (And the supply of atmospheric carbon dioxide is increasing these days as we burn coal and oil.)

There is also a vast quantity of nitrogen in the air; much more than there is in the ocean. This is available to ocean life, too, at least indirectly, through the activity of nitrogen-fixing bacteria. These con-

Table 12

	PER CENT COMPOSITION OF OCEAN	PER CENT COMPOSITION OF COPEPOD	CONCENTRATION FACTOR
Carbon	0.0031	6.10	2,000.
Nitrogen	0.00008	1.52	19,000.
Chlorine	2.04	1.05	0.52
Sodium	1.09	0.54	0.50
Potassium	0.042	0.29	6.9
Sulfur	0.097	0.14	1.4
Phosphorus	0.000011	0.13	12,000.
Calcium	0.0024	0.04	16.5
Magnesium	0.13	0.03	0.23
Iron	0.000002	0.007	3,500.
Silicon	0.0004	0.007	17.
Bromine	0.0072	0.0009	0.12
Iodine	0.000005	0.0002	40.

vert gaseous nitrogen, which is itself unusable to higher forms of life, into nitrates, which are usable.

For these reasons, neither carbon nor nitrogen can ever be considered bottlenecks against the additional formation of total protoplasm. There is only a finite quantity of both, but long before life feels the pinch in either carbon or nitrogen, there is the shortage of iron and phosphorus to be continued.

And here phosphorus is four times as critical as iron. The copepod, of course, is only one type of life, but in general this pattern carries through. Phosphorus has the highest concentration factor; it is the first element to be used up. Life can multiply until all the phosphorus is gone and then there is an inexorable halt which nothing can prevent.

Even that much is only possible under favorable energy conditions. For it takes energy to concentrate the phosphorus and iron of the ocean to the levels required by living tissue. For that matter, it takes energy to expel enough of the chlorine, sodium, magnesium, and bromine to bring their concentrations down to levels tolerated by living tissue. It also takes energy to convert the simple low-energy compounds of the ocean (even after appropriate concentration or thinning out) into the complicated high-energy compounds characteristic of living tissue.

The energy required is supplied by sunlight, which is inexhaustible in those places where it exists. Where it does exist, plant cells multiply and convert, by photosynthesis, the radiant energy of the sun into the chemical energy of carbohydrates, fats, and proteins. Animals (a form of life making up only a small portion of the total) obtain their energy by eating the plants' cells and metabolizing their tissue substance for the chemical energy it contains.

But sunlight only exists in the top 150 meters of the ocean. Below that, sunlight does not penetrate and plant cells do not grow. It is only in the top 150 meters (the "euphotic zone," from Greek words meaning "good light") that the energy supply itself is not a bottleneck and life can multiply in all its forms until all the phosphorus is used up.

And it does exactly that.

The inorganic phosphorus content of the surface ocean water itself is virtually zero. Just about all the phosphorus it contains is organic; that is, it is found either in the living cells or in the wastes and dead residues thereof.

What happens, then, in the euphotic zone, is a standoff. Animal life eats plant life, while plant life, using animal's wastes as a phosphorus source, grows to replace that portion of itself that has been eaten. The total volume of life is at a steady maximum.

Life below the euphotic zone depends for its existence on an organic rain from above. Animal organisms can swim downward out of the euphotic zone (and plant cells can be forced down by an unlucky current) and there they may be eaten by creatures that live in the sub-euphotic zone regularly.

Also, dead remnants of life drift downward. They are gobbled up by the animal life of the depths (no plant life below the euphotic zone) which in turn contribute, after death, to a continuous drizzle that moves always lower down. In the long run, this perpetually renewed drizzle supports all life down to the very floor of the abyss.

Below the euphotic zone, it is energy, not phosphorus, that is the bottleneck; energy in the form of the organic compounds of this drizzle, which animals can feed on (in addition to each other, of course) and which they can metabolize for energy. Below the euphotic zone, then, there is less life than is necessary to incorporate all the phosphorus of the surroundings. There are, therefore, inorganic phosphorus compounds (phosphates) remaining in the deep ocean water itself.

The organic drizzle represents a loss of phosphorus to the euphotic zone, for dead tissue and animal wastes are rich in that element. If there were nothing to counteract this transfer of phosphorus from the euphotic zone to the depths, the quantity of life in the euphotic zone would inexorably decline along with the concentration of phosphorus and eventually blink out.

Fortunately, there is circulation between the depths and the surface of the ocean. There is an upwelling of water from the abyss, rich in phosphorus, which replaces the phosphorus lost in the organic drizzle. This upwelling is greatest in cold waters such as those of the Antarctic and the North Atlantic. There the chill and heavy surface waters sink and are replaced from the depths. There, consequently, the euphotic zone is richest in phosphorus and can support the greatest concentration of life. (Giant whales, which require a great deal of food for maintenance, for that reason congregate in the Antarctic and North Atlantic. No fools they.)

On the other hand, the warm and light surface waters of the warm areas of Earth remain tenaciously afloat and are not directly replaced by the colder and heavier waters from the depths. They must depend on surface currents from the cold North and South to replenish their phosphorus content. This second-hand supply of phosphorus has already been plundered by the life-forms that reached it first, so the ocean life of the tropics is less rich than that of the colder zones. In warm, land-locked portions of the ocean, such as the Mediterranean Sea, which are relatively sheltered from the phosphorus-relief of even the cold surface current, the ocean life is still less rich.

On the whole, though, there is a balance everywhere in the oceans, and, again on the whole, it is the concentration of phosphorus, life's bottleneck, that dictates the nature of the balance.

The situation with respect to land-based life has special points of interest. Land life is a latecomer to the scene and is still, quantitatively speaking, a minor offshoot of the ocean. Something like 85 per cent of all living matter lives in water; only 15 per cent on land. It is only the fact that *Homo sapiens* happens to live on land that makes us give terrestrial environment the undue attention it receives.

On land, as you would expect of life-forms that had evolved in water, the real bottleneck is the water itself, which no longer surrounds and permeates the life-forms. Land life has cut down on its use of hydrogen and oxygen in consequence. Whereas hydrogen and oxygen together make up about 90 per cent of a copepod, it makes up only about 86 per cent of a land plant like alfalfa, and only 72 per cent of a land animal such as man.

The cut-down is not remarkably extensive, however, and if a territory receives insufficient water, life-forms are sparse regardless of what elements the soil might contain.

Granting the needed water, bottlenecks must next be sought for among elements other than oxygen or hydrogen. Carbon and nitrogen are eliminated on land for the same reason they were eliminated in the ocean. The atmospheric supply of nitrogen, thanks to nitrogen-fixing bacteria, is ample, and the carbon supply is fleshed out by the atmospheric carbon dioxide.

That leaves elements other than hydrogen, oxygen, carbon, and nitrogen. Leaving those four out, the remaining elements must all be derived from the minerals of the soil, ultimately. For these we can set up Table 13, comparing the percentage composition of Earth's crust with that of an example of terrestrial plant life, such as alfalfa. (On land, as in the sea, plant life predominates quantitatively, and animal life is absolutely dependent upon it. Whatever element is life's bottleneck for plants is therefore the bottleneck for animals as well.)

Table 13

	PER CENT COMPOSITION OF SOIL	PER CENT COMPOSITION OF ALFALFA	CONCENTRATION FACTOR
Phosphorus	0.12	0.706	5.9
Calcium	3.63	0.58	0.16
Potassium	2.59	0.17	0.066
Sulfur	0.052	0.104	2.0
Magnesium	2.09	0.082	0.039
Chlorine	0.048	0.070	1.5
Iron	5.00	0.0027	0.0005
Boron	0.0010	0.0007	0.70
Manganese	0.10	0.00036	0.0036
Zinc	0.0080	0.00035	0.044
Copper	0.0070	0.00025	0.036
Molybdenum	0.00023	0.00010	0.43
Iodine	0.00003	0.0000025	0.08
Cobalt	0.0040	0.0000010	0.00025

In some respects, the concentration factors in Table 13 are not as good as they appear. Comparing them with those in Table 12 would make it seem that by and large, soil is so much more concentrated with the various essential elements than is the ocean that life on land ought to outstrip life in the ocean by far.

Nevertheless, the fact is that elements contained in solid minerals are useless and unavailable to plant life, and consequently, in the long run, to animal life as well. The plant lives on the substances it can extract from solution in the water contained in the soil.

Since the minerals of the soil are relatively insoluble, the watery solution is a thin one indeed and concentration factors are actually very high. That is one reason why land-based life is actually sparser than sea life despite the greater apparent concentration of minerals on land than in the sea.

Furthermore, the material in the soil is not spread evenly. One region may have adequate quantities, let us say, of zinc or copper because of some local deposit, while a neighboring region may be deficient in both and another neighboring region may have a poisonous excess of both. Any element can represent a local bottleneck to life. This is one reason that, even given plenty of sun and rain, one section of land may be less fertile than another.

To be sure, there is an extremely slow soil-homogenizing factor in the land erosion that goes on over the ages, bringing materials from mountaintops to valleys; in the buckling of strata and the scraping of glaciers and the upraising of mountains. In the very long run, then, local deficiencies and excesses don't matter. It is the over-all concentration factor that matters, and there, on land as on the sea, phosphorus is the bottleneck.

Man can take a hand, of course. He can, to the limits imposed by his technology, make up for deficiencies without waiting for geologic processes. He can transfer water from points of excess (with the ocean as the basic source) to points of deficiency. He can do the same for nitrogen (with the air as the basic source) or for calcium or phosphorus.

In doing this, man is, in a way, trying to homogenize the soil and make it more evenly fertile. He is not raising the maximum potential of fertility. The maximum mass of protoplasm which the land can support, like the maximum that the sea can support, is dictated by the phosphorus content. Phosphorus, on both land and sea, has the highest concentration factor; on both land and sea it is life's bottleneck.

Just as there is a standoff in the euphotic zone, so there is a standoff on land. The rain comes down, dissolves tiny quantities of soil, and on this solution, plants grow until all the phosphorus they can grab has been incorporated into their substance. Animals eat the plants and, in the process of living, excrete phosphorus-containing wastes upon which plant life can feed, grow, and replace the amount of itself which animals have eaten.

And just as there is a drizzle out of the euphotic zone of the ocean, so there is a drizzle out of the land. Some of the dissolved materials in the soil inevitably escape the waiting rootlets and are carried by the seeping soil water to brooks and rivers and eventually to the sea.

Any one river in any one second doesn't transfer much in the way of dissolved substances from land to ocean, but all the rivers together pour 9,000 cubic miles of water into the oceans each year, and in that amount of water, even a very thin solution amounts to a lot of dissolved material.

The loss of phosphorus, since that is life's bottleneck, is most serious, and it is estimated that 3,500,000 tons of phosphorus are washed from the land into the sea by the rivers each year. Since phosphorus makes up roughly 1 per cent of living matter, that means that the potential maximum amount of land-based protoplasm deceases each year by 350,-000,000 tons.

Of course, there may be some device of retransfer from sea to land, just as in the case of the ocean there is a retransfer of phosphorus from the depths to the surface.

One type of retransfer of phosphorus from sea to land involves bird droppings. Some sea birds live on fish and nest on land. Their dropping are rich in phosphorus (derived from fish, which get it from the ocean) and these sea-derived droppings cover their nesting grounds by the ton. This material, called guano, is a valuable commodity since it is an excellent fertilizer just because of its phosphorus content.

However, the phosphorus returned to land in this fashion represents only 3 per cent or less of that washed out to sea. The rest is not returned!

Nor does the phosphorus washed into the sea remain dissolved there. If it did, life in the sea would gradually multiply as life on land diminished, and the total protoplasmic mass on Earth would remain unchanged. Unfortunately the ocean is already holding all it can of the largely insoluble phosphates. New phosphorus washed into the sea simply precipitates as sediment at the sea bottom.

Of course, over geologic periods, the uplifting of sea bottoms exposes new phosphorus-rich soil to start cycles of land fertility over again. At the present moment, though, this long-range view won't help us. With an increasing population, we need increased fertility of the soil just to stay even, and steadily decreasing fertility could spell disaster.

Especially when man is deliberately accelerating the rate at which phosphorus is being lost to the sea.

This is where the new villain comes in. In all advanced regions of Earth (and more and more regions are becoming advanced) internal plumbing is coming into fashion. Elaborate sewer pipes lace cities and phosphorus-rich wastes are carefully and thoroughly flushed into the ocean.

And so soil fertility declines even faster and cannot be replaced by the chemical industry, since more and more of the most necessary chemical, phosphorus, will be at the ocean's bottom, where man himself helped put it and from where man has no way of retrieving it as yet.

Naturally, I am not suggesting that we abandon plumbing and sewers. I am used to sanitation myself and have no real affection for things like typhoid fever and cholera which go along with lack of it.

I am suggesting, though, that while we try to cope with the inevitable disappearance of coal, oil, wood, space between people, and other things that are vanishing as population and per capita power requirements mount recklessly each year, we had better add the problem of disappearing phosphorus to the list, and do what we can to encourage sewage disposal units which process it as fertilizer rather than dump it as waste—or to mine the ocean-floor.

We may be able to substitute nuclear power for coal power, and plastics for wood, and yeast for meat, and friendliness for isolation—but for phosphorus there is neither substitute nor replacement.

11

THE EGG AND WEE

Every once in a while, you will come across some remarks pointing up how much more compact the human brain is than is any electronic computer.

It is true that the human brain is a marvel of compactness in comparison to man-made thinking machines, but it is my feeling that this is not because of any fundamental difference in the nature of the mechanism of brain action as compared with that of computer action. Rather, I have the feeling that the difference is a matter of the size of the components involved.

The human cerebral cortex, it is estimated, is made up of 10,000,-000,000 nerve cells.[1] In comparison, the first modern electronic computer, ENIAC, had about 20,000 switching devices. I don't know how many the latest computers have, but I am quite certain they do not begin to approach a content of ten billion.

The marvel, then, is not so much the brain as the cell. Not only is

---

[1] *I should say, to be more accurate, 10,000,000,000 neurons plus 90,000,000,000 glial cells.*

the cell considerably smaller than any man-made unit incorporated into a machine, but it is far more flexible than any man-made unit. In addition to acting as an electronic switch or amplifier (or whatever it does in the brain), it is a complete chemical factory.

Furthermore, cells need not aggregate in fearfully large numbers in order to make up an organism. To be sure, the average man may contain 50,000,000,000,000 (fifty trillion) cells and the largest whale as many as 100,000,000,000,000,000 (a hundred quadrillion) cells, but these are exceptional. The smallest shrew contains only 7,000,000,000 cells, and small invertebrate creatures contain even less. The smallest invertebrates are made up of only one hundred cells or so, and yet fulfill all the functions of a living organism.

As a matter of fact (and I'm sure you're ahead of me here), there are living organisms that possess all the basic abilities of life and are nevertheless composed of but a single cell.

If we are going to concern ourselves with compactness, then, let's consider the cell and ask ourselves the questions: How compact can a living structure be? How small can an object be and still have the capacity for life?

To begin: How large is a cell?

There is no one answer to that, for there are cells and cells, and some are larger than others. Almost all are microscopic, but some are so large as to be clearly, and even unavoidably, visible to the unaided eye. Just to push it to an extreme, it is possible for a cell to be larger than your head.

The giants of the cellular world are the various egg cells produced by animals. The human egg cell (or ovum), for instance, is the largest cell produced by the human body (either sex), and it is just visible to the naked eye. It is about the size of a pinhead.

In order to make the size quantitative and compare the human ovum in reasonable fashion with other cells both larger and smaller, let's pick out a convenient measuring unit. The inch or even the millimeter (which is approximately $\frac{1}{25.4}$ of an inch) is too large a unit for any cell except certain egg cells. Instead, therefore, I'm going to use the micron,[1] which equals a thousandth of a millimeter or $\frac{1}{25,400}$ of an inch. For volume, we will use a cubic micron, which is the volume

---

[1] *In 1962, when this article was written, I had not completely adopted the metric system. The "micron" is a poor term. Logically, it should be "micrometer."*

of a cube one micron long on each side. This is a very tiny unit of volume, as you will understand when I tell you that a cubic inch (which is something that is easy to visualize) contains over 16,000,-000,000,000 (sixteen trillion) cubic microns.

There are a third as many cubic microns in a cubic inch, then, as there are cells in a human body. That alone should tell us we have a unit of the right magnitude to handle cellular volumes.

Back to the egg cells then. The human ovum is a little sphere approximately 140 microns in diameter and therefore 70 microns in radius. Cubing 70 and multiplying the result by 4.18 (I will spare you both the rationale and the details of arithmetic manipulation), we find that the human ovum has a volume of a little over 1,400,000 cubic microns.

But the human ovum is by no means large for an egg cell. Creatures that lay eggs, birds in particular, do much better; and bird eggs, however large, are (to begin with, at least) single cells.

The largest egg ever laid by any bird was that of the extinct Aepyornis of Madagascar. This was also called the elephant bird, and may have given rise to the myth—so it is said—of the roc of the *Arabian Nights*. The roc was supposed to be so large that it could fly off with an elephant in one set of talons and a rhinoceros in the other. Its egg was the size of a house.

Actually, the Aepyornis was not quite that lyrically vast. It could not fly off with any animal, however small, for it could not fly at all. And its egg was considerably less than house-size. Nevertheless, the egg was nine and one-half inches wide and thirteen inches long and had a volume of two gallons, which is tremendous enough if you want to restrict yourself to the dullness of reality.

This is not only the largest egg ever laid by any bird, but it may be the largest ever laid by any creature, including the huge reptiles of the Mesozoic age, for the Aepyornis egg approached the maximum size that any egg, with a shell of calcium carbonate and without any internal struts or braces, can be expected to reach. If the Aepyornis egg is accepted as the largest egg, then it is also the largest cell of which there is any record.

To return to the here and now, the largest egg (and, therefore, cell) produced by any living creature is that of the ostrich. This is about six to seven inches in length and four to six inches in diameter; and, if you are interested, it takes forty minutes to hard-boil an ostrich egg.

In comparison, a large hen's egg is about one and three-quarter inches wide and two and a half inches long. The smallest egg laid by a bird is that of a species of hummingbird which produces an egg that is half an inch long.

Now let's put these figures, very roughly, into terms of volume in Table 14.

Table 14

EGG	VOLUME (IN CUBIC MICRONS)
Aepyornis	7,500,000,000,000,000
Ostrich	1,100,000,000,000,000
Hen	50,000,000,000,000
Hummingbird	400,000,000,000
Human being	1,400,000

As you see, the range in egg size is tremendous. Even the smallest bird egg is about 300,000 times as great in volume as the human ovum, whereas the largest bird egg is nearly 20,000 times as large as the smallest.

In other words, the Aepyornis egg compares to the hummingbird egg as the largest whale compares to a medium-sized dog; while the hummingbird egg, in turn, compares to the human ovum as the largest whale compares to a large rat.

And yet, even though the egg consists of but one cell, it is not the kind of cell we can consider typical. For one thing, scarcely any of it is alive. The eggshell certainly isn't alive and the white of the egg serves only as a water store. The yolk of the egg makes up the true cell and even that is almost entirely food supply.

If we really want to consider the size of cells, let's tackle those that contain a food supply only large enough to last them from day to day— cells that are largely protoplasm, in other words. These nonyolk cells range from the limits of visibility downward, just as egg cells range from the limits of visibility upward.

In fact, there is some overlapping. For instance, the amoeba, a simple free-living organism consisting of a single cell, has a diameter of about two hundred microns and a volume of 4,200,000 cubic microns. It is three times as voluminous as the human ovum.

The cells that make up multicellular organisms are considerably smaller, however. The various cells of the human body have volumes varying from 200 to 15,000 cubic microns. A typical liver cell, for in-stance, would have a volume of 1,750 cubic microns.

If we include cell-like bodies that are not quite complete cells, then we can reach smaller volumes. For instance, the human red blood cell, which is incomplete in that it lacks a cell nucleus, is considerably smaller than the ordinary cells of the human body. It has a volume of only 90 cubic microns.

Then, just as the female ovum is the largest cell produced by hu-man beings, the male spermatozoon is the smallest. The spermatozoon is mainly nucleus, and only half the nucleus at that. It has a volume of about 17 cubic microns.

This may make it seem to you that the cells making up a multi-cellular organism are simply too small to be individual and independ-ent fragments of life, and that in order to be free-living a cell must be unusually large. After all, an amoeba is 2,400 times as large as a liver cell, so perhaps in going from amoeba to liver cell, we have passed the limit of compactness that can be associated with independent life.

This is not so, however. Human cells cannot, to be sure, serve as individual organisms, but that is only because they are too specialized and *not* because they are too small. There are cells that serve as in-dependent organisms that are far smaller than the amoeba and even smaller than the human spermatozoon. These are the bacteria.

Even the largest bacterium has a volume of no more than 7 cubic microns, while the smallest have volumes down to 0.02 cubic microns. All this can be summarized in Table 15:

Table 15

NONYOLK CELL	VOLUME (IN CUBIC MICRONS)
Amoeba	4,200,000
Human liver cell	1,750
Human red blood cell	90
Human spermatozoon	17
Largest bacterium	7
Smallest bacterium	0.02

Again we have quite a range. A large one-celled organism such as the amoeba is to a small one-celled organism such as a midget bacterium, as the largest full-grown whale is to a half-grown specimen of the small-est variety of shrew. For that matter, the difference between the largest and smallest bacterium is that between a large elephant and a small boy.

Now, then, how on earth can the complexity of life be crammed into a tiny bacterium one two-hundred-millionth the size of a simple amoeba?

Again we are faced with a problem in compactness and we must pause to consider units. When we thought of a brain in terms of pounds, it was a small bit of tissue. When we thought of it in terms of cells, however, it became a tremendously complex assemblage of small units. In the same way, in considering cells, let's stop thinking in terms of cubic microns and start considering atoms and molecules.

A cubic micron of protoplasm contains about 40,000,000,000 molecules. Allowing for this, we can recast Table 15 in molecular terms in Table 16:

Table 16

CELL	NUMBER OF MOLECULES
Amoeba	170,000,000,000,000,000
Human liver cell	70,000,000,000,000
Human red blood cell	3,600,000,000,000
Human spermatozoon	680,000,000
Largest bacterium	280,000,000,000
Smallest bacterium	800,000,000

It would be tempting, at this point, to say that the molecule is the unit of the cell, as the cell is the unit of a multicellular organism. If we say that, we can go on to maintain that the amoeba is seventeen million times as complicated, molecularly speaking, as the human brain is, cellularly speaking. In that case, the compactness of the amoeba as a container for life becomes less surprising.

There is a catch, though. Almost all the molecules in protoplasm are water; simple little H_2O combinations. These are essential to life, goodness knows, but they serve largely as background. They are not *the* characteristic molecules of life. If we can point to any molecules as characteristic of life, they are the complex nitrogen-phosphorus macromolecules: the proteins, the nucleic acids and the phospholipids. These, together, make up only about one ten-thousandth of the molecules in living tissue.

(Now, I am *not* saying that these macromolecules make up only

$\frac{1}{10,000}$ of the *weight* of living tissue; only of the numbers of molecules. The macromolecules are individually much heavier than the water molecules. An average protein molecule, for instance, is some two thousand times as heavy as a water molecule. If a system consisted of two thousand water molecules and one average protein molecule, the *number* of protein molecules would only be $\frac{1}{2,001}$ of the total, but the *weight* of protein would be $\frac{1}{2}$ the total.)

Let's revise matters again as in Table 17:

Table 17

CELL	NITROGEN-PHOSPHORUS MACROMOLECULES
Amoeba	17,000,000,000,000
Human liver cell	7,000,000,000
Human red blood cell	360,000,000
Human spermatozoon	68,000,000
Largest bacterium	28,000,000
Smallest bacterium	80,000

We can say, then, that the average human body cell is indeed as complex, molecularly speaking, as the human brain, cellularly speaking. Bacteria, however, are markedly simpler than the brain, while the amoeba is markedly more complex.

Still, even the simplest bacterium grows and divides with great alacrity and there is nothing simple, from the chemical standpoint, about growing and dividing. That simplest bacterium, just visible under a good optical microscope, is a busy, self-contained and complex chemical laboratory.

But then, most of the 80,000 macromolecules in the smallest bacterium (say 50,000 at a guess) are enzymes, each of which can catalyze a particular chemical reaction. If there are 2,000 different chemical reactions constantly proceeding within a cell, each of which is necessary to growth and multiplication (this is another guess), then there are, on the average, 25 enzymes for each reaction.

A human factory in which 2,000 different machine operations are being conducted, with 25 men on each machine, would rightly be considered a most complex structure. Even the smallest bacterium is that complex.

We can approach this from another angle, too. About the turn of the century, biochemists began to realize that in addition to the obvious atomic components of living tissue (such as carbon, hydrogen, oxygen, nitrogen, sulfur, phosphorus, and so on) certain metals were required by the body in very small quantities.

As an example, consider two recent additions to the list of trace metals in the body, molybdenum and cobalt. The entire human body contains perhaps 18 milligrams of molybdenum and 12 milligrams of cobalt (roughly one two-thousandth of an ounce of each). Nevertheless, this quantity, while small, is absolutely essential. The body cannot exist without it.

To make this even more remarkable, the various trace minerals, including molybdenum and cobalt, seem to be essential to every cell. Divide up one two-thousandth of an ounce of these materials among the fifty trillion cells of the human body and what a miserably small trace of a trace is supplied each! *Surely*, the cells can do without.

But that is only if we persist in thinking in terms of ordinary weight units instead of in atoms. In the average cell, there are, very roughly speaking, some 40 molybdenum and cobalt atoms for every billion molecules. Let's, therefore, prepare Table 18:

Table 18

CELL	NUMBER OF MOLYBDENUM AND COBALT ATOMS
Amoeba	6,800,000,000
Human liver cell	2,800,000
Human red blood cell	144,000
Human spermatozoon	27,200
Largest bacterium	11,200
Smallest bacterium	32

(Mind you, the cells listed are not necessarily "average." I am quite certain that the liver cell contains more than an average share of these atoms and the red blood cell less than an average share; just as earlier, the spermatozoon undoubtedly contained more than an average share of macromolecules. However, I firmly refuse to quibble.)

As you see, the trace minerals are not so sparse after all. An amoeba possesses them by the billions of atoms and a human body cell by the millions. Even the larger bacteria possess them by the thousands.

The smallest bacteria, however, have only a couple of dozen of them, and this fits in well with my earlier conclusion that the tiniest bacterium may have, on the average, 25 enzymes for each reaction.

Cobalt and molybdenum (and the other trace metals) are essential because they are key bits of important enzymes. Allowing one atom per enzyme molecule, there are only a couple of dozen such molecules, all told, in the smallest bacterium.

But here we can sense that we are approaching a lower limit. The number of different enzymes is not likely to be distributed with perfect evenness. There will be more than a couple of dozen in some cases and less than a couple of dozen in others. Only one or two of the rarest of certain key enzymes may be present. If a cell had a volume of less than 0.02 cubic microns, the chances would be increasingly good that some key enzymes would find themselves jostled out altogether; with that, growth and multiplication would cease.

Therefore, it is reasonable to suppose that the smallest bacteria visible under a good optical microscope are actually the smallest bits of matter into which all the characteristic processes of life can be squeezed. Such bacteria represent, by this way of thinking, the limit of compactness as far as life is concerned.

But what about organisms still smaller than the smallest bacteria that, lacking some essential enzyme or enzymes, do not, under ordinary conditions, grow and multiply? Granted they are not independently alive, can they yet be considered as fully nonliving?

Before answering, consider that such tiny organisms (which we can call subcells) retain the potentiality of growth and multiplication. The potentiality can be made an actuality once the missing enzyme or enzymes are supplied, and these can only be supplied by a complete and living cell. A subcell, therefore, is an organism that possesses the ability to invade a cell and there, within the cell, to grow and multiply, utilizing the cell's enzymatic equipment to flesh out its own shortcomings.

The largest of the subcells are the rickettsiae, named for an American pathologist, Howard Taylor Ricketts, who, in 1909, discovered that insects were the transmitting agents of Rocky Mountain spotted fever, a disease produced by such subcells. He died the next year of typhus fever, catching it in the course of his researches on that disease, also transmitted by insects. He was thirty-nine at the time of his death; and his reward for giving his life for the good of man is, as you might expect, oblivion.

The smaller rickettsiae fade off into the viruses (there is no sharp dividing line) and the smaller viruses lap over, in size, the genes, which are found in the nuclei of cells and which, in their viruslike structure, carry genetic information.

Now, in considering the subcells, let's abandon the cubic micron as a measure of volume, because if we don't we will run into tiny decimals. Instead, let's use the "cubic millimicron." The millimicron[2] is $\frac{1}{1,000}$ of a micron. A cubic millimicron is, therefore, $\frac{1}{1,000}$ times $\frac{1}{1,000}$ times $\frac{1}{1,000}$, or one-billionth of a cubic micron.

In other words, the smallest bacterium, with a volume of 0.02 cubic microns, can also be said to have a volume of 20,000,000 cubic millimicrons. Now we can prepare Table 19, dealing with subcell volumes:

Table 19

SUBCELL	VOLUME (IN CUBIC MILLIMICRONS)
Typhus fever rickettsia	54,000,000
Cowpox virus	5,600,000
Influenza virus	800,000
Bacteriophage	520,000
Tobacco mosaic virus	50,000
Gene	40,000
Yellow-fever virus	5,600
Hoof-and-mouth virus	700

The range of subcells is huge. The largest rickettsia is nearly three times the size of the smallest bacterium. (It is not size alone that makes an organism a subcell; it is the absence of at least one essential enzyme.) The smallest subcell, on the other hand, is only $\frac{1}{3,500}$ as large as the smallest bacterium. The largest subcell is to the smallest one as the largest whale is to the average dog.

As one slides down the scale of subcells, the number of molecules decreases. Naturally, the nitrogen-phosphorus macromolecules don't disappear entirely, for life, however distantly potential, is impossible (in the form we know) without them. The very smallest subcells consist of nothing more than a very few of these macromolecules; only the bare essentials of life, so to speak, stripped of all superfluity.

The number of atoms, however, is still sizable. A cubic millimicron will hold several hundred atoms if they were packed with the greatest possible compactness, but of course, in living tissue, they are not.

[2] *Worse and worse! The proper term for "millimicron" in the metric system is "nanometer."*

Thus, the tobacco mosaic virus has a molecular weight of 40,000,000 and the atoms in living tissue have an atomic weight that averages about 8. (All but the hydrogen atom have atomic weights that are well above 8, but the numerous hydrogen atoms, each with an atomic weight of 1, pulls the average far down.)

This means there are roughly 5,000,000 atoms in a tobacco mosaic virus particle, or just about 100 atoms per cubic millimicron. We can therefore prepare Table 20, a new version of Table 19:

Table 20

SUBCELL	NUMBER OF ATOMS
Typhus fever rickettsia	5,400,000,000
Cowpox virus	560,000,000
Influenza virus	80,000,000
Bacteriophage	52,000,000
Tobacco mosaic virus	5,000,000
Gene	4,000,000
Yellow-fever virus	560,000
Hoof-and-mouth virus	70,000

It would seem, then, that the barest essentials of life can be packed into as few as 70,000 atoms. Below that level, we find ordinary protein molecules, definitely nonliving. Some protein molecules (definitely nonliving) actually run to more than 70,000 atoms, but the average such molecule contains 5,000 to 10,000 atoms.

Let's consider 70,000 atoms, then, as the "minimum life unit." Since an average human cell contains macromolecules possessing a total number of atoms at least half a billion times as large as the minimum life unit, and since the cerebral cortex of man contains ten billion such cells, it is not at all surprising that our brain is what it is.

In fact, the great and awesome wonder is that mankind, less than ten thousand years after inventing civilization, has managed to put to-gether a mere few thousand excessively simple units and build com-puters that do as well as they do.

Imagine what would happen if we could make up units containing half a billion working parts, and then use ten billion of those units to design a computer. Why, we would have something that would make the human brain look like a wet firecracker.

Present company excepted, of course!

12

THAT'S LIFE!

My son is fiendishly interested in outer space.[1] This is entirely without reference to his father's occupation, concerning which he is possessed of complete apathy. Anyway, in honor of this interest of his, we once bought a recording of a humorous skit entitled "The Astronaut" (which was soon worn so thin as the result of repeated playings, that the needle delivered both sides simultaneously).

At one point in this recording, the interviewer asks the astronaut whether he expects to find life on Mars, and the astronaut answers thoughtfully, "Maybe. . . . If I land on Saturday night."

Which brings us face to face with the question of what, exactly, do we mean by life. And we don't have to go to Mars to be faced with a dilemma. There is room for heated arguments right here on earth.

We all know, or think we know, purely on the basis of intuition, what life is. We know that we are alive and that a dead man isn't, that an oyster is alive and a rock isn't. We are also quite confident that such diverse things as sea anemones, gorillas, chestnut trees,

[1] At least, he was when this article was written, twelve years ago.

sponges, mosses, tapeworms, and chipmunks are all alive—except when they're dead.

The difficulty arises when we try to take this intuitive knowledge and fit it into words, and this is what I am going to try to do in this chapter.

There is more than one fashion in which we can construct a definition. For instance, we can make a functional definition, or we can make a structural one.

Thus, a child might say: "A house is something to live in" (functional). Or he might say: "A house is made of brick" (structural).

Neither definition is satisfactory since a tent is something to live in and yet is not ordinarily considered a house, while a wall may be made of brick and yet not be a house.

Combining the two types of definitions may leave it imperfect even so, but it will represent an improvement. Thus "a house is something made of brick in which people live" at once eliminates tents and walls. (It also eliminates frame houses, to say nothing of brick houses that are owned by families who have just left for a month's vacation in the mountains.)

This line of reasoning has an application to definitions involving the concept of life. For instance, when I went to school, the definition I saw most often was functional and went something like this: "A living organism is characterized by the ability to sense its environment and respond appropriately, to ingest food, digest it, absorb it, assimilate it, break its substance down and utilize the energy so derived, excrete wastes, grow and reproduce." (When I refer to this later in the chapter, I shall signify the list by "sense its environment, etc.," to save wear and tear on my typewriter ribbon and your retina.)

There was always a question, though, as to whether this was really an exclusive definition. Inanimate objects could imitate these functions if we wanted to argue subtly enough. Crystals grow, for instance, and if we consider the saturated solution to be its food, we certainly might make out a case for absorption and assimilation. Fires can be said to digest their fuel and to leave wastes behind, and they certainly grow and reproduce. Then, too, very simple robots have already been constructed that can imitate all these functions of life (except growth and reproduction) by means of a photocell and wheels.

I tried to define life functionally in another fashion in a book entitled *Life and Energy* (Doubleday, 1962). I introduced thermodynamic concepts and said: "A living organism is characterized by the ability to effect a temporary and local decrease in entropy."

As it stands, however, this definition is perfectly terrible, for a refrigerator can also bring about a temporary and local decrease in entropy; for example, every time it cools a quart of milk. However, as I shall explain later in the chapter, I don't let this statement stand unmodified.

What we need, clearly, is to introduce something structural into the definition, but can we? All forms of life, however diverse in appearance, have functions in common. They all sense their environment, etc., which is why a functional definition can be set up so easily. But do they all have any structure in common? The mere fact that I use the clause "however diverse in appearance" would indicate they do not.

That, however, is only true if we were to rely on the diversity of ordinary appearance as visible to (if you will excuse the expression) the naked eye. But suppose we clothe the eye in an appropriate lens?

Back in 1665, an English scientist, Robert Hooke, published a book in which he described his researches with a microscope. As part of his research, he studied a thin section of cork and found it to be riddled with tiny rectangular holes. He named the holes "cells," this meaning any small room and therefore being a graphically appropriate word.

But cork is dead tissue even when it occurs on a living tree. Over the next century and a half, microscopists studied living tissue, or, at least, tissue that was alive until they prepared it for study. They found that such tissue was also marked off into tiny compartments. The name "cell" was kept for those even though, in living tissue, such compartments were no longer empty holes but were, to all appearances, filled with matter.

It wasn't until the 1830s, though, that accumulating evidence made it possible for two German biologists, Matthias Jakob Schleiden and Theodor Schwann, to present the world with the generalization that all living organisms were made up of cells.

Here, then, is a structural definition: "A living organism is made up of cells."

However, such a definition, although it sounds good, cannot be reversed. You cannot say that an object composed of cells is living, since a dead man is made up of cells just as surely as a living man is, except that the cells of a dead man are dead.

And it does no good to amend the definition by saying that a living organism is composed of living cells, because that is arguing in a circle. Besides, in an organism that is freshly dead, many cells are still alive. Perhaps even the vast majority are—yet the organism is dead.

We can do better, as in the case of the definition of the house, if we include both structural and functional aspects in the definition and say: "A living organism is made up of cells *and* is characterized by the ability to sense its environment, etc."

Here is a definition that includes all the diverse types of organisms we intuitively recognize as living, and excludes everything else, such as crystals, river deltas, flames, robots, and abstractions which can be said to mimic the functions we associate with life—simply because these latter objects do not consist of cells. The definition also excludes dead remnants of once-living objects (however freshly dead) because such dead objects, while constructed of cells, do not perform the functions we associate with life.

I referred in passing, some paragraphs ago, to "living cells." What does that mean?

The definition of a living organism as I have just presented it says that it is made up of cells, but does that imply that the cells themselves are alive? Can we argue that all parts of a living body are necessarily alive and that cells therefore must be alive, as long as they are part of a living organism?

This is clearly a mistaken argument. Hair is not alive, though it is growing on your body. Your skin is covered with a layer of cells that are quite dead by any reasonable criterion though they are part of a living organism.

If we are going to decide whether cells are alive, we can't allow it to depend secondarily on a definition of a living organism. We must apply the necessary criteria of life to the cell itself and ask whether it can sense its environment, etc., and meet, at least, the functional definition of a living thing.

At once the answer is No. Many cells clearly lack one or more of the vital functional abilities of living things. The cells of our nervous system, for instance, cannot reproduce. We are born with the total number of nerve cells we will ever have; any change thereafter can only be for the worse, for a nerve cell that loses its function cannot be replaced.

To be more general, none of our cells, if detached from its neighbors and set up in business for itself, can long survive to fulfill its functions.

And yet there are different cells among those of our body that can, in themselves, perform each of the functions associated with life. Some cells can sense their environment; others respond appropriately; some supervise digestion; others absorb; all cells assimilate and produce and use energy; some cells grow and reproduce continually throughout life even after the organism as a whole has ceased to grow and reproduce. In short, the functions of a living organism are, in a sense, the sum of the functions of the cells making it up.

We can say then: "A living cell is one that contributes in some active fashion to the functioning of the organism of which it is a part." This raises the question of what we mean by "some active fashion," but I will leave that to your intuition and say only that the definition is intended to eliminate the problem of the dead cells of the skin, which serve our body only by being there as protection, and not by doing anything actively. Furthermore, a cell may continue its accustomed activities for a limited time after the death of the organisms, and then we can speak of living cells in a dead body.

But there is still an important point to make. We now have two different definitions, one for a living cell and one for a living organism.

THEODOR SCHWANN

Schwann was born in Neuss, Rhenish Prussia, on December 7, 1810, and when he was only twenty-four years old made his first important discovery. It was known at the time that the stomach juices contained hydrochloric acid and that this was quite capable of breaking up, or digesting, foodstuffs. In 1834, however, Schwann prepared extracts of the glandular lining of the stomach and showed that, mixed with acid, it had a far greater meat-dissolving power than acid alone. He had discovered "pepsin," the first digestive enzyme from animal tissue to be brought to the attention of science.

He is best known, however, for his elaboration, in 1839, of the cell theory. In its simplest form, this is the statement that all living things are made up of cells or of material formed by cells, and that each cell contains certain essential components such as a nucleus and a sur-

rounding membrane. (He also coined the term "metabolism" as representing the over-all chemical changes taking place in living tissue.)

Schwann pointed out that plants and animals alike were formed out of cells, that eggs were cells distorted by the presence of yolk, that eggs grew and developed by dividing and redividing so that the developing organism consisted of more and more cells, but always of cells. He even discovered one particular kind of cells, those that made up the nerve sheaths, and they are still called "Schwann cells."

In the last forty years of his life, Schwann did nothing to match his activities in the one decade of the 1830s, but we can scarcely complain about that. He died in Cologne, Rhenish Prussia, on January 11, 1882.

That means that a cell of a human being is not alive in the same sense that an entire human being is alive. And this makes sense at that, for though the functions of a human being may be viewed as the sum of the functions of his cells, the life of a human being is still more than the sum of the life of his cells.

If you can imagine all the cells of the human body alive in isolation and put together at random, you know that no human being will result. A human being consists not only of something material (cells), but of something rather abstract as well (a specific cell organization). It is quite possible to end human life by destroying the organization while scarcely touching any of the cells themselves.

But I am talking about human cells—is this true for the cells of other organisms as well? Yes, it is; at least, for any reasonably complex organism.

However, as one descends to simpler and simpler organisms, the factor of cellular organization becomes progressively less important. That is, disruption of organization can become more and more extensive without actually putting an end to life. We can replace a lost fingernail, but a lobster can replace a lost limb. A starfish can be cut into sizable chunks and each piece will grow back the remainder, while a sponge can be divided into separate cells which will then reclump and reorganize. At no point, however, is organization of zero importance, as long as an organism consists of more than one cell.

But organisms made up of a single cell do indeed exist, having first been discovered by a Dutch microscopist, Anton van Leeuwenhoek, at the same time that Hooke was discovering cells. A one-celled organism, such as an amoeba, fulfills all the functional requirements of a living organism, in that it can sense its environment, etc. And yet it does not meet the structural portion of the definition, for it is not composed of cells. It *is* a cell.

So we can modify the definition: "A living organism is made up of one or more cells and is characterized by the ability to sense its environment, etc."

It follows, then, that cell organization is not an absolute requirement for *all* types of living organisms. Only the existence of the cell itself seems to be required for the existence of a living thing.

For this reason, it grew popular in the nineteenth century to say that the cell was the "unit of life," and for biologists to devote more and more of their effort toward an understanding of the cell.

But now we can raise the question as to whether the cell actually is the irreducible unit of life or whether something still simpler exists that will serve in that respect.

First, what is a cell? Roughly speaking, we can speak of it as an object that contains at least three parts. First, it possesses a thin membrane that marks it off from the outside universe. Second, it possesses a small internal structure called a nucleus. Third, between membrane and nucleus lies the cytoplasm.

To be sure, there are human cells (such as those of the heart) that run together and are not properly separated by membranes. There are also human cells, such as the red blood corpuscles, that have no nuclei. These are, however, highly specialized cells of a multicellular organism which, in isolation, we cannot consider living organisms.

For those cells that are truly living organisms, it remains true that the membrane, cytoplasm and nucleus are minimum essentials. Some particularly simple one-celled organisms appear to lack nuclei, the bacteria and the blue-green algae being examples. These cells, however, contain "nuclear material"; that is, regions which react chemically as do the intact nuclei of more complicated cells. These simple cells still have nuclei then, but nuclei that are spread through the body of the cell rather than collected in one spot.

Is any one of these parts of the cell more essential than the other two? That may seem like asking which leg of a three-legged stool is more essential, since no cell can live without all three. Nevertheless there is evidence pointing to a gradation of importance. If an amoeba, for instance, is divided by means of a fine needle into two halves, one of which contains the intact nucleus, the half with the nucleus can recover, survive, grow and reproduce normally. The half without the nucleus may carry on the functions of life for a short while, but cannot grow or reproduce.

Furthermore, when a cell divides, it goes through a complicated series of changes that particularly involve small structures called chromosomes, which lie within the nucleus. This is true whether a cell is an organism in itself or is merely part of a larger organism.

The changes in which the chromosomes are involved include a key step, one in which each chromosome induces the formation of another like itself. This is called "replication," for the chromosome has produced a replica of itself. No cell ever divides without replication taking place. As the nineteenth century drew to an end, the suspicion began to stir in biologists that as the cell was the key to the organism, so the chromosome was the key to the cell.

We can help matters along if we turn once again to the structural definition. After all, our definition of a living organism is both functional and structural as far as multicellular organisms are concerned. They are composed of cells. For a one-celled organism, the definition becomes purely functional, for there is nothing to say what a single cell is composed of.

To clarify that point, we can descend to the molecular level. A cell contains numerous types of molecules, some of which are also to be found in inanimate nature and which are, therefore, however characteristic of living organisms, not characteristic *only* of living organisms. (Water is an example.)

Yet there are molecules that are to be found only in living cells or in material that was once part of a living cell or, at the very least, had been formed by a living cell. The most characteristic of these (for reasons I won't go into here) are the various protein molecules. No form of life exists, no single cell, however simple or however complicated, that does not contain protein.

Proteins satisfy a variety of functions. Some merely make up part of the substratum of the body, forming major components of skin, hair, cartilage, tendons, ligaments, and so on. Other proteins, however, are intimately concerned with the actual chemical workings of the cell; they catalyze the thousands of reactions that go on. These proteins (called enzymes) are, we cannot help but intuitively feel, close to the chemical essence of life.

In fact, I can now return to my book *Life and Energy*, from which I quoted an unsatisfactory definition of the living organism near the beginning of the chapter, and can explain how I amended the definition to make it satisfactory, thus: "A living organism is characterized by the ability to effect a temporary and local decrease in entropy by means of enzyme-catalyzed reactions." Here is a definition that is both functional (it effects an entropy decrease) and structural (by means of enzymes).

Now, this definition *does not involve cells*. It applies as truly to a multicellular as to a unicellular organism, and it accurately marks off those systems we intuitively recognize as alive from those we do not.

This new definition would make it seem that it is not the cell so much that is the unit of life, but the enzymes within the cell. However, if enzymes can only be formed within cells and by cells, the distinction is a purely academic one. Unless, that is, we can pin down the manufacture of enzymes to something more specific than the cell as a whole.

In recent decades it has become quite obvious that the thousands of different enzymes present in each cell (one for each of the thousands of different chemical reactions that are continually proceeding within the cell) are formed under the supervision of the chromosomes.

Shifting to the chromosomes then, and remaining on the molecular level, I must explain that the chromosomes are composed of a series of giant protein molecules of a variety called nucleoprotein, because each consists of a protein portion and a nucleic acid portion. The nucleic acid portion is quite different from the protein in structure.

Nucleic acid is so named because it was found, originally, in the nucleus. Since the first days, it has also been found in the cytoplasm, but it keeps its original name. There are two forms of nucleic acid, with complicated names that are abbreviated DNA and RNA. DNA is found only in the nucleus and makes up a major portion of the chromosomes. RNA is found chiefly in the cytoplasm, though a small quantity is also present in the nucleus.

Research in the 1950s has shown that it is not merely the chromosomes but the DNA content thereof (with an assist from RNA) that supervises the synthesis of specific enzymes. Through those enzymes, the nucleic acids of the cell might be said to supervise the chemical activity of the cell and to be therefore, in control of all the functions we associate with living organisms.

But though nucleic acids control the functions of living organisms, can they themselves be considered "living"? When this question arose earlier in the chapter in connection with cells, I wasn't satisfied that a cell was truly alive until it could be shown that a single cell could serve as an organism in itself. Similarly, we can't consider nucleic acids to be alive until and unless we can show that a nucleic acid molecule can serve as an organism in itself.

Let's go back in time again.

Back in the 1880s, the French biochemist Louis Pasteur while studying hydrophobia, tried to isolate the germ of the disease. Twenty

years earlier, you see, he had evolved the "germ theory of disease," which stated that all infectious diseases were caused and transmitted by microorganisms. Hydrophobia was certainly infectious, but where was the microorganism?

Pasteur had two choices. He could abandon his theory or he could

LOUIS PASTEUR

Pasteur, born in Dôle, France, on December 27, 1822, was not a remarkably good student as a boy. He was interested in painting and did moderately well in mathematics, but in chemistry (in which he was eventually to be a first-magnitude star) he received the mark of "mediocre."

He was turned from his ambition to be a professor of the fine arts by attending the lectures of the French chemist Jean Dumas. Thanks to Dumas' encouragement, Pasteur embarked on the path of chemical research with the most astounding results.

In 1848 Pasteur noted that certain tartrates crystallized in two fashions, one the mirror image of the other. He separated the crystals and in this way demonstrated the existence of geometrical and optical isomers, something that led to an understanding of the actual shapes of molecules.

In 1854 Pasteur was asked to tackle a problem that was destroying France's wine industry. Wine was turning sour as it fermented and financial losses were enormous. Pasteur studied the fermenting wine under the microscope and found that samples from those vats that were souring contained yeast cells of abnormal shape. He recommended gentle heating of wine to kill the yeast as soon as the wine fermented. It worked and he thus introduced the technique of "pasteurization."

Tackling the dying silkworms that were reducing the French silk industry to a shambles, Pasteur found, in 1865, abnormal microorganisms there too, and recommended the destruction of all caterpillars and mulberry leaves showing them. He had already demonstrated that microorganisms only developed from other microorganisms and now he saw the manner in which they could be transferred, as parasites, from host to host. He put forth the "germ theory of disease," the greatest single medical advance in history.

Through that Pasteur and others began the conquest of contagious disease, and in a century, doubled the life expectancy of human beings. Pasteur died near Paris on September 28, 1895.

introduce an *ad hoc* amendment (that is, one designed for no other purpose than to explain away a specific difficulty). Ordinarily the introduction of *ad hoc* amendments is a poor procedure, but a genius can get away with it. Pasteur suggested that the germ of hydrophobia existed, but was too small to be seen in a microscope.

Pasteur was right.

Another disease studied at the time, by botanists, was tobacco mosaic disease, one in which the leaves of tobacco plants were mottled into a mosaic. The juice from a diseased leaf would infect a healthy leaf, so by Pasteur's theory, a germ should exist. None, however, could be found here, either.

In 1892, a Russian bacteriologist, Dmitri Ivanovski, ran some of the juice of a diseased leaf through a porcelain filter that was so fine that no bacterium, not even the smallest, could pass through. The juice that did get through was still capable of passing on the disease. The infectious agent was therefore called a "filtrable virus." ("Virus" simply means "poison," so a "filtrable virus" is a poison that passes through a filter.)

Other diseases, including hydrophobia, were found to be transmitted by filtrable viruses. The nature of these viruses, however, was unknown until 1931, when an English bacteriologist, William J. Elford, designed a filter fine enough to trap the virus. In this way the virus, though smaller by far than even the smallest cells, proved to be larger by far than most molecules.

Well, then, was the virus particle (whatever its nature) a living organism? It infects cells so it must somehow sense their presence and respond appropriately. It must feed on their substance, absorb, assimilate, make use of energy, grow and reproduce. And yet the virus particle certainly did not consist of cells as they were then known. The whole problem of the nature of life was thrown into confusion in the 1930s, although the problem had been clarified by the cell theory in the 1830s.

In 1935, the American biochemist Wendell Meredith Stanley actually succeeded in crystallizing the tobacco mosaic virus, and this seemed to be a forceful argument against life. Even after the virus had been crystallized it remained infective, and how can anything living survive crystallization, for goodness' sake? Crystals were objects associated only with nonliving things.

Actually, this argument is worthless. Nothing alive can be crystallized because, until viruses were discovered, nothing alive was simple enough to be crystallized. But viruses were simpler than any cellular form of life and there was no reason in the world to suppose that the non-crystallization rule ought to apply to them.

Once enough tobacco mosaic virus was purified and brought together by crystallization, it could be tested chemically, and it was found by two British biochemists, Frederick C. Bawden and Norman W. Pirie, to be a nucleoprotein. It was 94 per cent protein and 6 per cent RNA.

Since then, without exception, all viruses that have been analyzed have proved to be nucleoprotein. Some contain DNA, some RNA, some both—but none are completely lacking in nucleic acid.

Furthermore, when a virus infects a cell, it is the nucleic acid portion that actually enters the cell, while the protein portion remains outside. There is now every reason to think that the protein is merely a nonliving shell about the nucleic acid, which is itself the key portion of the virus. Naked nucleic acid molecules have even been prepared from viruses and, in themselves, have remained slightly infective.

It certainly looks as though in the virus we have found our example of a nucleic acid molecule that in itself and by itself behaves as a living organism.

Suppose we say, then: "A living organism is characterized by the possession of at least one molecule of nucleic acid capable of replication." This definition is both structural (the nucleic acid) and functional (the replication). It includes not only all cellular life, but all viruses as well; and it excludes all things else.

To be sure, there are arguments against this. Some feel that the virus is not a true example of a living organism because it cannot perform its function until it is inside the cell. Within the cell, and only within the cell, does it supervise enzyme action and bring about the synthesis of specific enzymes and other proteins. It does this by making use of the cell's chemical machinery, including its enzymes. Outside the cell, the virus performs none of the functions we associate with life. The cell, therefore, so this argument goes, is still the unit of life.

I do not see the force of this argument. To be sure, the virus requires a cell in order to perform certain of its functions, but its life outside the cell is not wholly static. It must actively penetrate the cell, and must do that without the help of the cell itself. This is an example of at least one action characteristic of life (the equivalent of ingestion of food, somewhat inside-out) that it performs all by itself.

Then, even if we admit that the virus makes use of cellular machinery for some of its functions, so does a tapeworm make use of our cellular machinery for some of its functions. The virus, like the tapeworm, is a parasite, but happens to be a more complete one. Shall we draw an artificial line and say that the tapeworm is a living organism and the virus is not?

Furthermore, all organisms, parasites or not, are as dependent upon some factor of the outer world as viruses are. We ourselves, for instance, could not live for more than a few minutes if our access to oxygen were cut off. Is that any reason to suppose that we are not living organisms but that it is the oxygen that is really the living organism? Why, therefore, put the necessary outside cell (for the virus) in a category different from that of the necessary outside oxygen (for us)?

Nor is there anything crucial in the fact that the virus makes use of enzymes that are not its own. Let me explain this point by analogy.

Consider the woodcutter chopping down a tree with an ax. He can't do it without an ax, and yet we never think of a woodcutter as a man-ax combination. A woodcutter is a man and the ax is merely the woodcutter's tool. Similarly, a nucleic acid may not be able to perform its actions without enzymes, but the enzymes are merely its tool, while the nucleic acid is the thing itself.

Furthermore, when a woodcutter is in action, chopping down a tree, the ax may be his or it may be stolen. This may make him an honest man or a thief, respectively, but in either case, he is a woodcutter in action. In the same way, a virus performing its functions is a living organism whether the enzymes it uses are its own or not.

As far as I am concerned, therefore, my definition of living organisms in terms of their nucleic acid content is a valid one.

It's necessary to remember, of course, that a living organism is more than its nucleic acid content, just as it is more than its cellular content. As I said earlier in the chapter, a living organism consists not only of separate parts, but of those parts in appropriate organization.

There are some biologists who deplore the intense concentration on DNA in contemporary biological and biochemical research. They feel that organization is being neglected in favor of a study of the parts alone, and I must admit there is some justification to this.

Nevertheless, I also feel that we will never understand the organization until we have a thorough understanding of the parts being organized, and it is my hope that when the DNA molecule is laid out plain for all to see, in all its detail and complexity, many of the current mysteries of life will fall neatly into place—organization and all.

13

NOT AS WE KNOW IT

Even unpleasant experiences can be inspiring.

For instance, my children once conned me into taking them to a monster-movie they had seen advertised on TV.

"It's *science fiction*," they explained. They don't exactly know what science fiction is, but they have gathered it's something daddy writes, so the argument is considered very powerful.

I tried to explain that it wasn't science fiction by *my* definition, but although I had logic on my side, they had decibels on theirs.

So I joined a two-block line consisting of every kid for miles around with an occasional grown-up who spent his time miserably pretending he was waiting for a bus and would leave momentarily. It was a typical early spring day in New England—nasty drizzle whipped into needle-spray by a howling east wind—and we inched slowly forward.

Finally, when we were within six feet of the ticket-sellers and I, personally, within six inches of pneumonia, my guardian angel smiled and I had my narrow escape. They hung up the SOLD OUT sign.

I said, with a merry laugh, "Oh, what a dirty shame," and drove my howlingly indignant children home.

Anyway, it got me to thinking about the lack of imagination in movieland's monsters. Their only attributes are their bigness and destructiveness. They include big apes, big octopuses (or is the word "octopodes"?), big eagles, big spiders, big amoebae. In a way, that is all Hollywood needs, I suppose. This alone suffices to drag in huge crowds of vociferous human larvae, for to be big and destructive is the secret dream of every red-blooded little boy and girl in the world.

What, however, is mere size to the true *aficionado?* What we want is real variety. When the cautious astronomer speaks of life on other worlds with the qualification "life-as-we-know-it," we become impatient.

What about life-not-as-we-know-it?

Well, that's what I want to discuss in this chapter.

To begin with, we have to decide what life-as-we-know-it means. Certainly life-as-we-know-it is infinitely various. It flies, runs, leaps, crawls, walks, hops, swims, and just sits. It is green, red, yellow, pink, dead white, and varicolored. It glows and does not glow, eats and does not eat. It is boned, shelled, plated, and soft; has limbs, tentacles, or no appendages at all; is hairy, scaly, feathery, leafy, spiny, and bare.

If we're going to lump it all as life-as-we-know-it, we'll have to find out something it all has in common. We might say it is all composed of cells, except that this is not so. The virus, an important life form to anyone who has ever had a cold, is not.

So we must strike beyond physiology and reach into chemistry, saying that all life is made up of a directing set of nucleic acid molecules which controls chemical reactions through the agency of proteins working in a watery medium.

There is more, almost infinitely more, to the details of life, but I am trying to strip it to a basic minimum. For life-as-we-know-it, water is the indispensable background against which the drama is played out, and nucleic acids and proteins are the featured players.

Hence any scientist, in evaluating the life possibilities on any particular world, instantly dismisses said world if it lacks water; or if it possesses water outside the liquid range, in the form of ice only or of steam only.

(You might wonder, by the way, why I don't include oxygen as a basic essential. I don't because it isn't. To be sure, it is the substance most characteristically involved in the mechanisms by which most life forms evolve energy, but it is not invariably involved. There are tissues in our body that can live temporarily in the absence of molecular oxygen, and there are microorganisms that can live indefinitely in the absence of oxygen. Life on Earth almost certainly developed in an oxygen-free atmosphere, and even today there are microorganisms that can live *only* in the absence of oxygen. No known life form on Earth, however, can live in the complete absence of water, or fails to contain both protein and nucleic acid.)

In order to discuss life-not-as-we-know-it, let's change either the background or the feature players. Background first!

Water is an amazing substance with a whole set of unusual properties which are ideal for life-as-we-know-it. So well fitted for life is it, in fact, that some people have seen in the nature of water a sure sign of Divine providence. This, however, is a false argument, since life has evolved to fit the watery medium in which it developed. Life fits water, rather than the reverse.

Can we imagine life evolving to fit some other liquid, then, one perhaps not too different from water? The obvious candidate is ammonia.

Ammonia is very like water in almost all ways. Whereas the water molecule is made up of an oxygen atom and two hydrogen atoms (H_2O) for an atomic weight of 18, the ammonia molecule is made up of a nitrogen atom and three hydrogen atoms (NH_3) for an atomic weight of 17. Liquid ammonia has almost as high a heat capacity as liquid water, almost as high a heat of evaporation, almost as high a versatility as a solvent, almost as high a tendency to liberate a hydrogen ion.

In fact, chemists have studied reactions proceeding in liquid ammonia and have found them to be quite analogous to those proceeding in water, so that an "ammonia chemistry" has been worked out in considerable detail.

Ammonia as a background to life is therefore quite conceivable—but not on Earth. The temperatures on Earth are such that ammonia exists as a gas. Its boiling point at atmospheric pressure is $-33.4°$ C ($-28°$ F) and its freezing point is $-77.7°$ C ($-108°$ F).

But other planets?

In 1931, the spectroscope revealed that the atmosphere of Jupiter, and, to a lesser extent, of Saturn, was loaded with ammonia. The notion arose at once of Jupiter being covered by huge ammonia oceans.

To be sure, Jupiter may have a temperature not higher than $-100°$ C ($-148°$. F), so that you might suppose the mass of ammonia upon it to exist as a solid, with atmospheric vapor in equilibrium. Too bad. If Jupiter were closer to the sun . . .

But wait! The boiling point I have given for ammonia is at atmospheric pressure—Earth's atmosphere. At higher pressures, the boiling point would rise, and if Jupiter's atmosphere is dense enough and deep enough, ammonia oceans might be possible after all.

An objection that might, however, be raised against the whole concept of an ammonia background for life, rests on the fact that living organisms are made up of unstable compounds that react quickly, subtly and variously. The proteins that are so characteristic of life-as-we-know-it must consequently be on the edge of instability. A slight rise in temperature and they break down.

A drop in temperature, on the other hand, might make protein molecules too stable. At temperatures near the freezing point of water, many forms of non-warm-blooded life become sluggish indeed. In an ammonia environment, with temperatures that are a hundred or so Celsius degrees lower than the freezing point of water, would not chemical reactions become too slow to support life?

The answer is twofold. In the first place, why is "slow" to be considered "too slow"? Why might there not be forms of life that live at slow motion compared to ourselves? Plants do.

A second and less trivial answer is that the protein structure of developing life adapted itself to the temperature by which it was surrounded. Had it adapted itself over the space of a billion years to liquid ammonia temperatures, protein structures might have been evolved that would be far too unstable to exist for more than a few minutes at liquid water temperatures, but are just stable enough to exist conveniently at liquid ammonia temperatures. These new forms would be just stable enough and unstable enough at low temperatures to support a fast-moving form of life.

Nor need we be concerned over the fact that we can't imagine what those structures might be. Suppose we were creatures who lived constantly at a temperature of a dull red heat (naturally with a chemistry fundamentally different from that we now have). Could we under those circumstances know anything about Earth-type proteins? Could we refrigerate vessels to a mere 25° C, form proteins and study them? Would we ever dream of doing so, unless we first discovered life forms utilizing them?

Anything else besides ammonia now?

Well, the truly common elements of the universe are hydrogen, helium, carbon, nitrogen, oxygen, and neon. We eliminate helium and neon because they are completely inert and take part in no reactions. In the presence of a vast preponderance of hydrogen throughout the universe, carbon, nitrogen, and oxygen would exist as hydrogenated compounds. In the case of oxygen, that would be water (H_2O), and in the case of nitrogen, that would be ammonia (NH_3). Both of these have been considered. That leaves carbon, which, when hydrogenated, forms methane (CH_4).

There is methane in the atmosphere of Jupiter and Saturn, along with ammonia; and, in the still more distant planets of Uranus and Neptune, methane is predominant, as ammonia is frozen out. This is because methane is liquid over a temperature range still lower than that of ammonia. It boils at $-161.6°$ C ($-259°$ F) and freezes at $-182.6°$ C ($-297°$ F) at atmospheric pressure.

Could we then consider methane as a possible background to life with the feature players being still more unstable forms of protein? Unfortunately, it's not that simple.

Ammonia and water are both polar compounds; that is, the electric charges in their molecules are unsymmetrically distributed. The electric charges in the methane molecule are symmetrically distributed, on the other hand, so it is a nonpolar compound.

Now, it so happens that a polar liquid will tend to dissolve polar substances but not nonpolar substances, while a nonpolar liquid will tend to dissolve nonpolar substances but not polar ones.

Thus water, which is polar, will dissolve salt and sugar, which are also polar, but will not dissolve fats or oils (lumped together as "lipids" by chemists), which are nonpolar. Hence the proverbial expression, "Oil and water do not mix."

On the other hand, methane, a nonpolar compound, will dissolve lipids but will not dissolve salt or sugar.

Proteins and nucleic acids are polar compounds and will not dissolve in methane. In fact, it is difficult to conceive of any structure that would jibe with our notions of what a protein or nucleic acid ought to be that would dissolve in methane.

If we are to consider methane, then, as a background for life, we must change the feature players.

To do so, let's take a look at protein and nucleic acid and ask ourselves what it is about them that makes them essential for life.

Well, for one thing, they are giant molecules, capable of almost infinite variety in structure and therefore potentially possessed of the versatility required as the basis of an almost infinitely varying life.

Is there no other form of molecule that can be as large and complex as proteins and nucleic acids and that can be nonpolar, hence soluble in methane, as well? The most common nonpolar compounds associated with life are the lipids, so we might ask if it is possible for there to exist lipids of giant molecular size.

Such giant lipid molecules are not only possible; they actually exist. Brain tissue, in particular, contains giant lipid molecules of complex structure (and of unknown function). There are large "lipoproteins" and "proteolipids" here and there which are made up of both lipid portions and protein portions combined in a single large molecule. Man is but scratching the surface of lipid chemistry; the potentialities of the nonpolar molecule are greater than we have, until recent decades, realized.

Remember, too, that the biochemical evolution of Earth's life has centered about the polar medium of water. Had life developed in a nonpolar medium, such as that of methane, the same evolutionary forces might have endlessly proliferated lipid molecules into complex and delicately unstable forms that might then perform the functions we ordinarily associate with proteins and nucleic acids.

Working still further down on the temperature scale, we encounter the only common substances with a liquid range at temperatures below that of liquid methane. These are hydrogen, helium, and neon. Again, eliminating helium and neon, we are left with hydrogen, the most common substance of all. (Some astronomers think that Jupiter may be four-fifths hydrogen, with the rest mostly helium—in which case good-by ammonia oceans after all.)

Hydrogen is liquid between temperatures of $-253°$ C $(-423°$ F$)$ and $-259°$ C $(-434°$ F$)$, and no amount of pressure will raise its boiling point higher than $-240°$ C $(-400°$ F$)$. This range is only twenty to thirty Celsius degrees over absolute zero, so that hydrogen forms a conceivable background for the coldest level of life. Hydrogen is nonpolar, and again it would be some sort of lipid that would represent the featured player.

So far the entire discussion has turned on planets colder than the Earth. What about planets warmer?

To begin with, we must recognize that there is a sharp chemical division among planets. Three types exist in the solar system and presumably in the universe as a whole.

On cold planets, molecular movements are slow, and even hydrogen and helium (the lightest and therefore the nimblest of all substances) are slow-moving enough to be retained by a planet in the process of formation. Since hydrogen and helium together make up almost all of matter, this means that a large planet would be formed. Jupiter, Saturn, Uranus, and Neptune are the examples familiar to us.

On warmer planets, hydrogen and helium move quickly enough to escape. The more complex atoms, mere impurities in the overriding ocean of hydrogen and helium, are sufficient to form only small planets. The chief hydrogenated compound left behind is water, which is the highest-boiling compound of the methane-ammonia-water trio and which, besides, is most apt to form tight complexes with the silicates making up the solid crust of the planet.

Worlds like Mars, Earth, and Venus result. Here, ammonia and methane forms of life are impossible. Firstly, the temperatures are high enough to keep those compounds gaseous. Secondly, even if such planets went through a super-ice-age, long aeons after formation, in which temperatures dropped low enough to liquefy ammonia or methane, that would not help. There would be no ammonia or methane in quantities sufficient to support a world-girdling life form.

Imagine, next, a world still warmer than our medium trio: a world hot enough to lose even water. The familiar example is Mercury. It is a solid body of rock with little, if anything, in the way of hydrogen or hydrogen-containing compounds.

Does this eliminate any conceivable form of life that we can pin down to existing chemical mechanisms?

Not necessarily.

There are nonhydrogenous liquids, with ranges of temperature higher than that of water. The most common of these, on a cosmic scale, would be sulfur which, under one-atmosphere pressure, has a liquid range from 113° C (235° F) to 445° C (833° F); this would fit nicely into the temperature of Mercury's sunside.

But what kind of featured players could be expected against such a background?

So far all the complex molecular structures we have considered have been ordinary organic molecules; giant molecules, that is, made up chiefly of carbon and hydrogen, with oxygen and nitrogen as major "impurities" and sulfur and phosphorus as minor ones. The carbon and hydrogen alone would make up a nonpolar molecule; the oxygen and nitrogen add the polar qualities.

In a watery background (oxygen-hydrogen) one would expect the oxygen atoms of tissue components to outnumber the nitrogen atoms, and on Earth this is actually so. Against an ammonia background, I imagine nitrogen atoms would heavily outnumber oxygen atoms. The two subspecies of proteins and nucleic acids that result might be differentiated by an O or an N in parentheses, indicating which species of atom was the more numerous.

The lipids, featured against the methane and hydrogen backgrounds, are poor in both oxygen and nitrogen and are almost entirely carbon and hydrogen, which is why they are nonpolar.

But in a hot world like Mercury, none of these types of compounds could exist. No organic compound of the types most familiar to us, except for the very simplest, could long survive liquid sulfur temperatures. In fact, earthly proteins could not survive a temperature of 60° C for more than a few minutes.

How then to stabilize organic compounds? The first thought might be to substitute some other element for hydrogen, since hydrogen would, in any case, be in extremely short supply on hot worlds.

So let's consider hydrogen. The hydrogen atom is the smallest of all atoms and it can be squeezed into a molecular structure in places where other atoms will not fit. Any carbon chain, however intricate, can be plastered round and about with small hydrogen atoms to form "hydrocarbons." Any other atom, but one, would be too large.

And which is the "but one"? Well, an atom with chemical properties resembling those of hydrogen (at least as far as the capacity for taking part in particular molecular combinations is concerned) and one which is almost as small as the hydrogen atom, is that of fluorine. Unfortunately, fluorine is so active that chemists have always found it hard to deal with and have naturally turned to the investigation of tamer atomic species.

This changed during World War II. It was then necessary to work with uranium hexafluoride, for that was the only method of getting uranium into a compound that could be made gaseous without trouble. Uranium research had to continue (you know why), so fluorine had to be worked with, willy-nilly.

As a result, a whole group of "fluorocarbons," complex molecules made up of carbon and fluorine rather than carbon and hydrogen, were developed, and the basis laid for a kind of fluoro-organic chemistry.

To be sure, fluorocarbons are far more inert than the corresponding hydrocarbons (in fact, their peculiar value to industry lies in their inertness) and they do not seem to be in the least adaptable to the flexibility and versatility required by life forms.

However, the fluorocarbons so far developed are analogous to polyethylene or polystyrene among the hydro-organics. If we were to judge the potentialities of hydro-organics only from polyethylene, I doubt that we would easily conceive of proteins.

No one has yet, as far as I know, dealt with the problem of fluoroproteins or has even thought of dealing with it—but why not consider it? We can be quite certain that they would not be as active as ordinary proteins at ordinary temperatures. But on a Mercury-type planet, they would be at higher temperatures, and where hydro-organics would be destroyed altogether, fluoro-organics might well become just active enough to support life, particularly the fluoro-organics that life forms are likely to develop.

Such fluoro-organic-in-sulfur life depends, of course, on the assumption that on hot planets, fluorine, carbon, and sulfur would be present in enough quantities to make reasonably probable the development of life forms by random reaction over the life of a solar system. Each of these elements is moderately common in the universe, so the assumption is not an altogether bad one. But, just to be on the safe side, let's consider possible alternatives.

Suppose we abandon carbon as the major component of the giant molecules of life. Are there any other elements which have the almost unique property of carbon—that of being able to form long atomic chains and rings—so that giant molecules reflecting life's versatility can exist?

The atoms that come nearest to carbon in this respect are boron and silicon, boron lying just to the left of carbon on the periodic table (as usually presented) and silicon just beneath it. Of the two, however, boron is a rather rare element. Its participation in random reactions to produce life would be at so slow a rate, because of its low concentration in the planetary crust, that a boron-based life formed within a mere five billion years is of vanishingly small probability.

That leaves us with silicon, and there, at least, we are on firm ground. Mercury, or any hot planet, may be short on carbon, hydrogen, and fluorine, but it must be loaded with silicon and oxygen, for these are the major components of rocks. A hot planet which begins by lacking hydrogen and other light atoms and ends by lacking silicon and oxygen as well, just couldn't exist because there would be nothing left in enough quantity to make up more than a scattering of nickel-iron meteorites.

Silicon can form compounds analogous to the carbon chains. Hydrogen atoms tied to a silicon chain, rather than to a carbon chain, form the "silanes." Unfortunately, the silanes are less stable than the corresponding hydrocarbons and are even less likely to exist at high temperatures in the complex arrangements required of molecules making up living tissue.

Yet it remains a fact that silicon does indeed form complex chains in rocks and that those chains can easily withstand temperatures up to white heat. Here, however, we are not dealing with chains composed of silicon atoms only (Si-Si-Si-Si-Si) but of chains of silicon atoms alternating with oxygen atoms (Si-O-Si-O-Si).

It so happens that each silicon atom can latch on to four oxygen atoms, so you must imagine oxygen atoms attached to each silicon atom above and below, with these oxygen atoms being attached to other silicon atoms also, and so on. The result is a three-dimensional network, and an extremely stable one.

But once you begin with a silicon-oxygen chain, what if the silicon atom's capacity for hooking on to two additional atoms is filled not by more oxygen atoms but by carbon atoms, with, of course, hydrogen atoms attached? Such hybrid molecules, both silicon- and carbon-based, are the "silicones." These, too, have been developed chiefly during World War II and since, and are remarkable for their great stability and inertness.

Again, given greater complexity and high temperature, silicones might exhibit the activity and versatility necessary for life. Another possibility: Perhaps silicones may exist in which the carbon groups have fluorine atoms attached, rather than hydrogen atoms. Fluorosilicones would be the logical name for these, though, as far as I know—and I stand very ready to be corrected—none such have yet been studied.

Might there possibly be silicone or fluorosilicone life forms in which simple forms of this class of compound (which can remain liquid up to high temperatures) might be the background of life and complex forms the principal characters?

Here, then, in Table 21, is my list of life chemistries, spanning the temperature range from near red heat down to near absolute zero:

Table 21

1. Fluorosilicone in fluorosilicone
2. Fluorocarbon in sulfur
3.* Nucleic acid/protein (O) in water
4. Nucleic acid/protein (N) in ammonia
5. Lipid in methane
6. Lipid in hydrogen

Of this half dozen, the third only is life-as-we-know-it. Lest you miss it, I've marked it with an asterisk.

This, of course, does not exhaust the imagination, for science-fiction writers have postulated metal beings living on nuclear energy, vaporous beings living in gases, energy beings living in stars, mental beings living in space, indescribable beings living in hyperspace, and so on.

It does, however, seem to include the most likely forms that life can take as a purely chemical phenomenon based on the common atoms of the universe.

Thus, when we go out into space there may be more to meet us than we expect. I would look forward not only to our extraterrestrial brothers who share life-as-we-know-it. I would hope also for an occasional cousin among the life-not-as-we-know-it possibilities.

In fact, I think we ought to prefer our cousins. Competition may be keen, even overkeen, with our brothers, for we may well grasp at one another's planets; but there need only be friendship with our hot-world and cold-world cousins, for we dovetail neatly. Each stellar system might pleasantly support all the varieties, each on its own planet, and each planet useless to and undesired by any other variety.

How easy it would be to observe the Tenth Commandment then!

PART V

GEOCHEMISTRY

14

RECIPE FOR A PLANET

Slowly, American scientists (and, I believe, Soviet scientists, independently) are making ready to drill a hole through the Earth's crust to reach the layer beneath.[1]

This projected "Mohole" (and I'll explain the name, for those of you who happen not to know, later on) will, if it succeeds, bring us the first direct information concerning any portion of our planet other than the very rind. This is exciting for several reasons, one of which is that it will lower the highblood pressure of many a geologist who for years has had to watch man make ready to go millions of miles out in space while totally unable to penetrate more than a few miles below Earth's outer surface. And there is something annoying (if you're a geologist) in the thought that mankind will certainly feel, in its own corporeal hands, a sample of the surface of Mars long before it can possibly feel a sample from the central regions of our own planet.

And yet we ought to look at the bright side. The wonder is not that

---

[1] *No, they're not. This article was written in 1961 and the project has been abandoned since.*

we are so helpless in the face of some thousands of miles of rigid impenetrability. Naturally, we're helpless. The wonder is that, being so helpless, we have nevertheless deduced as much information about the interior of the Earth as we have.

Of course, there are parts of the Earth that we can see and feel and which we can subject to our various instruments. Once modern chemistry was established by Lavoisier, there was no serious trouble in analyzing the composition of the atmosphere and of the oceans ("hydrosphere"). The former is, essentially, a mixture of oxygen, nitrogen, and argon gases in the ratio, roughly, of 78:21:1. The latter is, essentially, a 3 per cent water solution of sodium chloride, with some added impurities.

In addition, the outermost portions of the solid matter forming the body of the planet ("lithosphere") are within reach. This, however, presents a new problem. Atmosphere and hydrosphere are homogeneous; that is, if you analyze any small portion of it, you have the composition of the whole. The solid earth itself is heterogeneous; one portion is not necessarily at all like another, which is why we have diamonds in Kimberley and gold in the Klondike but nothing but cheap dirt and some dowdy crabgrass in my backyard.

This means that in order to find out the overall composition of the soil and rocks, analyses have to be run on different samples from different areas of the world, and some sort of average must be taken after estimating that there is so much of this kind of rock on Earth and so much of that. Various geologists have done this and come up with estimates that agree fairly well.

A typical such estimate is presented in Table 22, with the major elements of the Earth's crust presented in order of percentage by weight:

Table 22

Oxygen	46.60
Silicon	27.72
Aluminum	8.13
Iron	5.00
Calcium	3.63
Sodium	2.83
Potassium	2.59
Magnesium	2.09

The eight elements make up just over 98.5 per cent of the weight of the Earth's outermost layer. The remaining ninety-odd elements can be considered as trace impurities (very important ones in some cases, to be sure, since included among them are elements such as carbon, hydrogen, nitrogen, and phosphorus, which are essential to life).

Now none of the elements in the list occurs free; all are found in combination—with each other, naturally, since there is little else to combine with. The most obvious combination is that between silicon and oxygen (which together make up three-fourths of the weight of Earth's outermost layer) to form silicon dioxide or silica. Quartz is an example of relatively pure silica, while flint is less pure. Sand is weathered silica. In combination with the other six elements listed (all metals), silicon and oxygen form silicates.

In brief, then, the solid Earth's reachable portion can be looked on as a mixture of silica and silicates, with all else chicken feed, at least in terms of quantity.

The distribution of elements in the Earth's crust seems lopsided, but, as it happens, when we calculate that distribution by weight, as in the above list, we are making it as unlopsided as possible. Let us suppose that we estimated the composition by numbers of atoms instead of by weight.

Of the eight major elements of the Earth's crust, oxygen happens to have the lightest atom. That means that a fixed weight of oxygen will contain 1.75 times as many atoms as that same weight of silicon, 2.5 times as many as the same weight of potassium, 3.5 times as many as the same weight of iron.

If you count by atom, then, it turns out that of every 100 atoms in the Earth's crust, 62.5 are oxygen. To put it another way, pick up a handful of soil and chances are that five out of eight of the atoms you are holding are oxygen.

But matters are even more lopsided than that. In forming compounds with silicon and with the six major metals, the oxygen atom accepts electrons; the others all donate them. When an atom accepts electrons, those additional electrons take up orbits (to use the term loosely) on the very outskirts of the atom, swooping far out from the nucleus, which holds them rather weakly. Since the radius of an anion (i.e. an atom plus one or more electrons in excess) extends to the farthest electronic orbit, the oxygen anion is larger than the oxygen atom proper.

On the other hand, an element that gives up an electron or two, gives up just those outermost ones that are most weakly held. The remaining electrons cluster relatively closely about the nucleus, and the radius of such a cation (i.e. an atom with a deficiency of one or more electrons) is smaller than that of the original atom.

The result is that the oxygen anion has a radius of 1.40 Angstroms (an Ångstrom unit is a hundred-millionth of a centimeter), while the silicon cation has a radius of 0.42 Å and the iron cation has one of 0.74 Å. This despite the fact that the silicon and iron cations are each considerably more massive than the relatively light oxygen anion.

The volume of any sphere varies as the cube of the radius, so that the discrepancy in radii among the ions becomes much magnified in the volume itself. For instance, the volume of the oxygen anion is about 11.5 cubic Å, while the volume of the iron cation is only 2.1 cubic Å and the volume of the silicon cation is only 0.4 cubic Å.

Allowing for the greater number of oxygen atoms and the greater volume of the individual oxygen anion, it turns out that no less than 93.77 per cent of the *volume* of the Earth's crust is taken up by oxygen. The solid Earth on which we walk is a well-packed set of oxygen anions, crowded closely together, with the small cations of the other seven elements tucked in here and there in the interstices.

That goes for the Rock of Gibraltar, too—just a heap of oxygen and little more.

All this data deals, of course, with those portions of the lithosphere which we can gouge out and pulverize and put through the analytic mill. What about those portions we can't test? Mankind has dug some three miles deep into the crust in pursuit of gold, and a couple of miles deeper in chase of oil, but these are highly localized pinpricks. All but the surface is beyond our ken and may even be forever beyond said ken.

The lazy man's solution to the problem is to suppose that, in general, the surface of the Earth's crust is a fair representation of its interior and that the planet is the same through and through as it is on the surface.

Unfortunately for those seeking a simple answer, this isn't so on the face of it. If the Earth as a whole were as rich in uranium and thorium as the crust is, our planet would melt with the quantity of radioactive heat radiated. Just the fact that the Earth is solid, then, shows that those two elements peter out a short distance below Earth's skin, and proves that in one small way, at least, heterogeneity with depth exists.

Furthermore, the predominant rock of the continental masses is granite, while the predominant rock of the ocean bottoms would seem to be basalt. Granite is richer in aluminum than basalt is, and poorer in magnesium, so that some geologists have visualized the Earth's crust as consisting of comparatively light continental blocks rich in aluminum silicate ("sial") floating on a comparatively heavy underpinning rich in magnesium silicate ("sima"), with the Earth's water supply filling the gaps between the sial blocks.

This is probably an oversimplification, but it still brings up the notion that the composition of the Earth changes with depth. And yet, so far, it is only the metals that are involved. There is nothing in what I have said that seems to affect the point of silicon and oxygen preponderance. Whatever the change in detail, the Earth might still be a silicate ball in essence—one big globe of rock, in other words.

The first actual information, as opposed to pure guesswork, that was obtained about the Earth's deep interior, came when Henry Cavendish first determined the mass of the planet in 1798. The volume was known since ancient Greek times; and dividing Cavendish's mass by the volume gave the overall density of the Earth as 5.52 grams per cubic centimeter.

Now, the density of the Earth's crust is about 2.8 grams per cubic centimeter, and this means that the density must rise with increasing depth. In fact, it must rise well beyond the 5.52 mark to make up for the lower-than-average density of the surface layers.

This, in itself, is no blow to the rock-ball theory of Earth's structure, because, obviously, pressure must increase with depth. The weight of overlying layers of rocks must compress the lower layers more and more down to the center where, it is estimated, the pressure is something like 50,000,000 pounds to the square inch. The same rock which had a density of 2.8 grams per cubic centimeter on the surface might conceivably be squeezed to perdition and a density of, say, 12 grams per cubic centimeter at the Earth's center.

A more direct line of attack on the deep interior involves the study of earthquakes. By 1900, the Earth was beginning to be girdled by a network of seismographs equipped to study the vibrations set up in the body of the planet by the quakes.

Two main types of earthquake waves are produced, the P (or primary) waves and the S (or secondary) waves. The P waves are longitudinal alternating bands of compression and expansion, like sound waves. The S waves are transverse and have the ordinary snakelike wiggles we associate with waves. The P waves travel more rapidly than do the S waves, and are the first to arrive at a station. The further a station from the actual earthquake, the greater the lag in time before the S waves arrive. Three stations working together can use such time-lag data to spot, with great precision, the point of origin ("epicenter") of the earthquake.

Knowing the location of both the earthquake and the station, it is possible to plot the general path taken by the waves through the body of the Earth. The greater the distance between the earthquake and the receiving station, the more deeply would the arriving waves have penetrated the body of the Earth. If the Earth were uniformly dense and rigid, the time taken by the waves to arrive would be in proportion to the distance traveled.

Actually, however, the density of the Earth's substance is not uniform with depth, and neither is the rigidity of the material. Laboratory experiments on various rocks have shown how the velocity of the two types of waves vary with differing degrees of density and rigidity under various temperatures and pressures. Such data can be extrapolated to levels of temperature and pressure that are encountered in the depths of the Earth but cannot be duplicated in the laboratory. This is admittedly a risky business—extrapolation always is—but geologists feel confident they can translate the actual velocity of earthquake waves at a given depth into the density of the rock at that depth.

It turns out, then, that the density of the Earth does increase fairly slowly and smoothly from the 2.8 grams per cubic centimeter at the surface to about 5.9 grams per cubic centimeter at a depth of some 2,150 miles.

And then, suddenly, there is a sharp break. The fact that this is so can again be told from the behavior of the earthquake waves. As the waves progress through regions of increasing density with depth, then decreasing density as they approach the surface again, they change direction and are refracted, just as light waves are refracted on passing through changing densities of air. As long as the density change is gradual, the direction change is gradual too, and forms a smooth curve. This is exactly what happens as long as the wave in its progress does not penetrate more than 2,150 miles beneath the surface.

Imagine a station, then, at such a distance from the quake itself that the resulting waves have penetrated to that depth. All stations between itself and the quake receive waves, too, at times varying with the distance, with penetrations of less than 2,150 miles.

A station somewhat further from the earthquake epicenter might expect, reasonably enough, to receive waves that have penetrated deeper than 2,150 miles, but it receives no waves at all. Yet stations still further away by a thousand miles or so may receive waves clearly, albeit after a longer interval.

In short, there is an area (the "shadow zone") over the Earth's surface, forming a kind of doughnut-shape at a fixed distance from each particular earthquake epicenter, in which no waves are felt. The interpretation is that there is a sudden sharp change in direction in any wave penetrating past the 2,150 mile mark, so that it is sent beyond the shadow zone. The only reason for such a sudden sharp change in direction would be a sudden sharp change in density.

Analysis of arrival times outside the shadow zone shows that this sharp rise in density is from 5.9 to 9.5 grams per cubic centimeter. Below 2,150 miles, the density continues to rise smoothly with increasing depth, reaching a value of about 12 grams per cubic centimeter at the center of the Earth.

All this refers to the P waves only. The S waves are even more dramatic in their behavior. When an S wave penetrates below 2,150 miles, it is not merely altered in direction—it is stopped altogether. The most logical explanation arises from the known fact that longitudinal waves such as the P waves can travel through the body of a liquid, while transverse waves, such as the S waves, cannot. Therefore, the regions of the Earth lower than 2,150 miles must be liquid.

On the basis of earthquake wave data, then, we must assume the Earth to consist of a liquid "core" about 1,800 miles in radius, which is surrounded by a solid "mantle" about 2,150 miles thick. The sharp division between these two major portions was first clearly demonstrated by the work of the American geologist Beno Gutenberg in 1914, and is therefore called the "Gutenberg discontinuity."

In 1909, a Serbian geologist named Andrija Mohorovicic (there are two little accent marks, different ones, over the two c's in his name which, for simplicity's sake, I will omit) discovered a sudden change in the velocity of earthquake waves on a line about 20 miles below the surface. This is called the "Mohorovicic discontinuity" and, because imagination boggles at the numbers of tonsils that would be thrown into disorder every time such a phrase was mouthed, it is becoming customary to say "Moho" instead. Moho is taken as the line separating the mantle below from the "crust" above.

Since those days, a detailed study of Moho shows that it is not at uniform depth. Under coastal land areas the depth *is* about 20 miles (it is 22 miles under New York, for instance), but under mountainous areas it dives lower—as low as 40 miles. (Since the crust is lighter than the mantle, you could say that mountains are mountains because the unusual concentration of light crust present causes the region to "float high.")

Conversely, Moho comes fairly close to the surface under some parts of the dense ocean bottom, which being comparatively heavy for crust, "float low." In some places, Moho is only 8 to 10 miles below sea level. This is particularly interesting because the ocean itself is 5 to 7 miles deep in spots and there is no problem in drilling through water. The actual thickness of solid material between ourselves and Moho can be as little as 3 miles if the right spot in the ocean is selected.

In one of those spots we propose to dig what you can now see must be called the "Mohole" (what else?) to reach the mantle. What's more, my own suggestion is that the ship carrying the equipment be named the "Moholder," but I guess that no one will pay any mind to me at all, with regard to this.

If we are going to consider the overall composition of the Earth, it is only necessary to deal with the core and the mantle. The core makes up only one-sixth of the Earth's volume, but because of its relatively high density it represents almost one-third of its mass. The remaining two-thirds is the mantle. The crust makes up only $\frac{1}{250}$ of the Earth's mass, and the hydrosphere and atmosphere are even more insignificant than that. We can therefore completely ignore the only parts of the Earth on which we have direct analytical data.

What about the mantle and core, then? The mantle differs from the crust only slightly in density and other properties, and everyone agrees that it must be essentially silicate in nature. Experiments in the laboratory show that a rock called olivine (a magnesium iron silicate) will, under high pressure, carry vibrations in the same range of velocities as that at which the mantle transmits earthquake waves. The general impression then is that the mantle differs from the crust in being much more homogeneous, richer in magnesium, and poorer in aluminum.

And the core? Still silicate, perhaps, but silicate which, at 2,150 miles, undergoes a sudden change in structure? In other words, may it not be that silicate is squeezed more and more tightly with mounting pressure until, at a certain point, something suddenly gives and all the atoms move into a far more compact arrangement? (This is analogous to the way in which the carbon atoms of graphite will move into the more compact arrangement of diamond if pressure and temperature are high enough.)

Some have proposed this, but there is no actual evidence that at the pressures and temperatures involved (which cannot be duplicated as yet in the laboratory) silicate will behave so.

The alternative is that there is a sudden change in chemical nature of the Earth's body, with the comparatively light silicate of the mantle giving way to some substance, heavier and liquid, that will make up the core.

But what material? If we arbitrarily restrict ourselves only to the elements common in the crust, the only substance that would be dense enough at deep-Earth pressures (and not too dense) and liquid at deep-Earth temperatures would be iron.

But isn't this pulling a rabbit out of that hat, too?

Not quite. There is still another line of evidence that, while terribly indirect, is very dramatic. The iron core was first suggested by a French geologist named Daubrée in 1866, a generation before earthquake data had pinpointed the existence of a core of some sort. His reasoning was based on the fact that so many meteorites consisted almost entirely of iron. This meant that pure iron could be expected to make up a portion of astronomical bodies. Hence, why not an iron core to Earth?

As a matter of fact, there are three kinds of meteorites: the "iron meteorites" already referred to; a group of much more common "stony meteorites" and a relatively rare group of "troilite meteorites." It is almost overwhelmingly tempting to suppose that these meteorites are remnants of an Earthlike planet (between Mars and Jupiter, where else?) that broke into fragments; that the stony meteorites are fragments of the mantle of that planet; the iron meteorites fragments of its core; and the troilite meteorites fragments of an intermediate zone at the bottom of its mantle.

If this is so, and most geologists seem to assume it is, then by analyzing the three sets of meteorites, we are, in effect, analyzing the Earth's mantle and core, at least roughly.

The stony meteorites have, on the whole, the composition in per cent by weight as shown in Table 23.

As you see, the stony meteorites are mainly a magnesium iron silicate,

Table 23

Oxygen	43.12
Silicon	21.61
Magnesium	16.62
Iron	13.23
Calcium	2.07
Aluminum	1.83

essentially an olivine. This, with calcium and aluminum as major impurities, comes to 98.5 per cent of the whole. Sodium and potassium, the other metals common in the crust, apparently peter out somewhat in the mantle. It's good to have them where we can reach them, though. They're useful elements and essential to life.

The composition of iron meteorites by per cent is as shown in Table 24:

Table 24

Iron	90.78
Nickel	8.59
Cobalt	0.63

Nothing else is present in significant quantity. Because of such analyses, the Earth's core is often referred to as the "nickel-iron core."

As for the composition of the troilite meteorites, it is as shown in Table 25:

Table 25

Iron	61.1
Sulfur	34.3
Nickel	2.9

This group consists, essentially, of iron sulfide[2] with a nickel sulfide impurity. Geologists therefore feel that the lowest portion of the Earth's mantle may well consist of a zone of iron sulfide making up perhaps one-twelfth of the Earth's total mass.

In order to get a notion of the overall composition of the Earth, un-

[2] *"Troilite" is the name given to an iron sulfide mineral similar to that found in these meteorites. It is named for an eighteenth-century Italian, Domenico Troili.*

der the assumption that its various major divisions correspond to the different varieties of meteorites, it is only necessary to get a properly weighted average of the meteoric data. Various geologists have made somewhat different assumptions as to the general composition of this or that part of the mantle and have come up with tables that differ in detail but agree in general.

Here, then, in Table 26, is one summary of the chemical composition of the whole Earth in percentages by weight:

Table 26

Iron	35.4
Oxygen	27.8
Magnesium	17.0
Silicon	12.6
Sulfur	2.7
Nickel	2.7

These six elements make up 98 per cent of the entire globe. If, however, the elements were listed not by weight but by atom number, the relatively light oxygen atoms would gain at the expense of the others and would move into first place. In fact, nearly half (47.2 per cent) of all the atoms in the Earth are oxygen.

It remains now to give the recipe for a planet such as ours; and I imagine it ought to run something like this, as it would appear in *Mother Stellar's Planetary Cookbook:*

"Weigh out roughly two septillion kilograms of iron, adding 10 per cent of nickel as stiffening. Mix well with four septillion kilograms of magnesium silicate, adding 5 per cent of sulfur to give it that tang, and small quantities of other elements to taste. (Use 'Mother Stellar's Elementary All-Spice' for best results.)

"Heat in a radioactive furnace until the mass is thoroughly melted and two mutually insoluble layers separate. (CAUTION. Do not heat too long, as prolonged heating will induce a desiccation that is not desirable.)

"Cool slowly till the crust hardens and a thin film of adhering gas and moisture appears. (If it does not appear, you have overheated.) Place in an orbit at a comfortable distance from a star and set to spinning. Then wait. In several billion years it will ferment at the surface. The fermented portion, called life, is considered the best part by connoisseurs."

15

NO MORE ICE AGES?

We all know that the radioactive ash resulting from the activities of nuclear power plants is dangerous and its disposal a problem to be brooded over. How different from those nice, decent, nonradioactive, old-fashioned coal-burning (or oil-burning) power plants. We can easily put ourselves into the position of the gentleman of the twenty-fifth century moaning thus for the good old days.

Except that the gentleman of the twenty-fifth century may well be sitting there cursing the good old days as he pushes his air conditioner up a notch and wishes that nuclear reactors—radioactive ash and all— had taken over a few centuries sooner than they did.

For coal and oil release an ash also and that ash is also puffed into the atmosphere. The ash of coal and oil isn't radioactive to be sure; it is only good old harmless carbon dioxide, which is already present in the atmosphere anyway.

It is only a minor constituent of the atmosphere, 0.04 per cent by weight, but this comes out into big numbers if all the air Earth has is lumped into the scale. The weight of our atmosphere is 5.70×10^{15}

tons, so the weight of the carbon dioxide in our atmosphere is 2.28×10^{12} (about two and a quarter trillion) tons.

That carbon dioxide, however, is subjected to some tremendous pushes and pulls. For instance, all plant life depends for existence on the consumption of atmospheric carbon dioxide. Using the energy of sunlight plus hydrogen atoms (obtained from water molecules), the plants convert the carbon dioxide to carbohydrate and then to all the other organic molecules necessary for the plant structure and chemistry.

Lump all the plant life by land and sea (especially the sea where the algae use up eight times as much carbon dioxide as all land plants put together) and a considerable amount of the gas is used up. Estimates for the carbon dioxide used up by plant life in one year vary from 60 to 200 billion tons. Even allowing the lower figure, it would seem that the carbon dioxide supply of the atmosphere would be used up in about thirty-six years. The larger figure will consume it in less than twelve years. Then all life comes to an end?

A fearful prospect, except that, of course, when an individual plant dies, bacteria attack the tissues and convert the carbon content back to carbon dioxide. And while plants live, they are at the mercy of marauding animals which do not utilize atmospheric carbon dioxide but get their energy supplies by tearing down what the plants have built up. They *form* carbon dioxide as the result of their life processes and exhale it back into the atmosphere.

So there is a carbon dioxide cycle, with plants using it up and animals and bacteria forming it again. If animals gain a temporary ascendancy, plant life is killed off at too rapid a rate and enough animals starve to allow plants a chance to revive. If they revive too far, animals multiply in the lush environment and cut the plants down once more. So there are minor oscillations which (if never allowed to oscillate too far in either direction, and so far—knock wood—they haven't) average out, in the long run, into perfect balance.

Well, not in perfect balance. There are leaks in both directions.

For instance, some dead plant tissues don't get consumed by bacteria but get covered by muck and mire and clamped down underground where, under heat and pressure, the organic molecules are slowly stripped of everything but carbon and hydrogen, and sometimes all the way to carbon only. Thus oil and coal are formed and the carbon atoms contained therein are withdrawn permanently (or for hundreds of millions of years anyway) from the carbon dioxide pool of the air.

Also, carbon dioxide may react with the inorganic rocks to form insoluble carbonates and may be removed more or less permanently in that way.

Balancing both leaks out of the atmosphere is new carbon dioxide leaking into the atmosphere as the result of volcanic action.

With leaks in *both* directions, there is the possibility of balance still. At the present time, in fact, there is such a balance. About 15 to 30 million tons of carbon dioxide are removed permanently from the atmosphere each year as coal or insoluble carbonate. The same amount is restored each year by volcanic action. (Notice that the inorganic contribution to the cycle is not more than 0.05 per cent of the biochemical contribution. Here's an example of the importance of life on a planetary scale.)

Still, did the leaks always balance? After all, there may have been periods in Earth's history when the leak in one direction or another became particularly prominent. There were long periods of time when coal formation proceeded at an unusually high rate. The trillions of tons of coal that are buried underground have all been withdrawn, however slowly, from the carbon dioxide pool of the air. Was that large-scale withdrawal replaced?

Again, during periods of mountain building, new rock is exposed to the atmosphere. Much more carbon dioxide than usual is used up in weathering and in the formation of insoluble carbonates. Is that carbon dioxide replaced?

On the other hand, there are periods of increased volcanic activity when more carbon dioxide is poured into the atmosphere than is true of most times.

Now then, does all this change the carbon dioxide content of the atmosphere from geologic era to geologic era? Probably yes, even if only slightly.

But if only slightly, does it matter? The answer to that is that some scientists think, yes, it matters a heck of a lot.

The major components of the atmosphere (oxygen and nitrogen) are, it seems, excellent transmitters of radiant energy over a broad stretch of wave lengths. The light rays of the Sun hit the air, pass through a hundred miles of it, hit the surface of the Earth, and are absorbed. The Earth heats up. The heated Earth radiates energy at night back into space, in the form of the far less energetic infra-red. This also passes through the atmosphere. The warmer Earth grows, the more

heat is radiated away at night. At some particular equilibrium temperature, the net loss of radiation by Earth at night equals that gained by
day so that, once that temperature (whatever it is) is reached, the Earth
as a whole neither warms nor cools with time (barring internal radioactivity). Of course, individual portions of it may warm and cool with
the seasons but this averages out, taken over the whole planetary surface.

Carbon dioxide, however, introduces a complication. It lets light rays
through as easily as do oxygen and nitrogen, but it absorbs infra-red
rather strongly. This means that Earth's nighttime radiation finds the
atmosphere partially opaque and some doesn't get through. The result
is that the equilibrium temperature must rise a few degrees to reach the
point where enough infra-red is forced out into space to balance the
Solar input. The Earth is warmer (on the whole) than it would be if
there were no carbon dioxide at all in the atmosphere. This warming
effect of carbon dioxide is called the "greenhouse effect."

If there were a period of increased weathering or coal formation, so
that the general carbon dioxide level of the atmosphere were to sink,
the greenhouse effect would decrease and the Earth's over-all temperature would drop. If volcanic action were to increase the carbon dioxide
level, the over-all temperature would rise.

A recent set of calculations indicate that if the present carbon dioxide
level should double, the over-all temperature of the Earth would rise by
3.6° C. If it were to halve, the temperature would drop 3.8° C.

Now to start an ice age going, you do not require a catastrophic
temperature drop. The drop need just be enough to allow a little more
snow to fall during the slightly colder winter than can be melted by the
succeeding, slightly cooler summer. Repeat this year after year and the
glaciers begin advancing. The chilled air and water drifting down from
the North make the summers cooler than ever and the process accelerates.

The amount of temperature drop below the present level required to
bring this about is not certainly known. Figures varying from a drop of
1.5 to 8° C have been suggested. Adopting a middle-of-the-road position, cutting the carbon dioxide of the atmosphere in half (from 0.04
to 0.02 per cent) would drop the temperature 3.8° C and that might
well be enough to start an ice age. Perhaps such a change was actually
the trigger of the ice ages that did happen.

A rise of 3 or 4°, on the other hand, would allow the slightly warmer
summers to melt just a little bit more ice than can be replaced by the

snows of the succeeding, slightly milder winters. The icecaps would melt and eventually disappear. There are currently 23 million cubic kilometers of ice in the world (mostly in Antarctica) and if this all melts, the volume of the oceans will increase by 1.7 per cent, the sea level would rise about 60 yards, and the coastal areas of the world would be flooded. (The Empire State Building would be in water to nearly the twentieth story.)

Obviously neither ice age nor world-wide tropic is desirable. Where we are is nice. Are we sure we are balanced or is there a slight trend one way or the other? Well, if there is, the trend ought to be so slight we need not worry for a million years—except for one thing.

Homo sapiens is throwing a monkey wrench into the machinery. We ourselves are upsetting the level by burning coal and oil in our, as aforesaid, nice, decent, nonradioactive, old-fashioned coal-burning (or oil-burning) power plants.

Until about 1900 the amount of carbon dioxide we formed in this manner was negligible. However, our industrialized twentieth century has been utilizing the "fossil fuels" in a logarithmically increasing fashion and the carbon dioxide that leaked out of the atmosphere over the space of a hundred million years of coal-forming is now being poured back into the atmosphere in a hundred million simultaneous puffs of smoke.

At the moment, we are adding 6 billion tons of carbon dioxide to the air each year (two hundred times as much as is being added by volcanic action and at least a fiftieth as much as is being added by life activity proper). And the rate is still increasing.

Even if we don't increase this rate, we will double the amount of carbon dioxide in the air (assuming there is no counteracting factor), raise the over-all temperature of the Earth 3.6° C, and make a healthy start at melting the icecaps *in toto* and drowning the coastal areas in a mere 350 years.

So much for our nice, decent, nonradioactive, old-fashioned coal-burning (or oil-burning) power plants.

Unless there is a counteracting factor. But is there?

Answer—maybe!

The first possibility is that as the atmospheric level of carbon dioxide goes up, plant life might luxuriate correspondingly, use the carbon di-

oxide faster, and bring the level down again. This would happen to begin with, probably. But then the natural interplay of life would balance
that. More plants alive means more plants dying and decaying. It also
means more animals to eat those plants. More decay and more animals
means more carbon dioxide produced. The level would go back up
again.

In other words, increasing the carbon dioxide of the air would speed
up the carbon dioxide cycle but would not introduce a corrective influence. If we increased the carbon dioxide content of the air, it would
stay increased, for all life processes could do.

But there is another factor. Leaving water vapor out of account, only
one of the normal components of the air, carbon dioxide, is appreciably
soluble in water. At $0°$ C, for instance, a milliliter (abbreviated ml) of
pure water will dissolve 0.0233 cubic centimeters (abbreviated cc) of
nitrogen, and 0.0489 cc of oxygen; but it will dissolve 1.713 cc of carbon
dioxide.

Now the oceans on Earth (which make up more than 98 per cent of
Earth's total water supply) contain a total of 1.37×10^{24} ml of an
aqueous salt solution. If this all held carbon dioxide at the rate of 1.713
cc per ml (so that the oceans fizzed like a planetful of soda water) there
would be 2.35×10^{24} cc of carbon dioxide in solution. That would
come, in weight, to 5.1×10^{15} tons, or about 2,250 times as much
carbon dioxide as there is in our entire atmosphere.

And actually this is a conservative estimate, since the solubility figures I gave are for pure water. This solubility goes up if the water is
made alkaline, and sea water is indeed somewhat alkaline.

If the ocean can dissolve so much carbon dioxide, it seems odd that
there remains any significant quantity of the gas in the atmosphere,
unless the ocean happens to be saturated with it. It is nowhere near
saturated, but the atmosphere retains the gas because the solution of
carbon dioxide depends on a number of local factors (temperature, pressure, acidity, salinity, the life processes of ocean-dwelling organisms,
etc.). Things are not as simple as though we put the oceans in a beaker
and bubbled the atmosphere through it, stirring vigorously all the while.

By actual measurement, it has been estimated that the total carbon
dioxide in the oceans is only 50 times that of the atmosphere.

Still, if this is the equilibrium, why shouldn't it be maintained when
mankind goes about pouring carbon dioxide into the atmosphere while

burning coal and oil. In other words, since 98 per cent of Earth's carbon dioxide is in the oceans, why shouldn't 98 per cent of Earth's *new* carbon dioxide go into the oceans.

If the ocean did in fact dissolve 98 per cent of new carbon dioxide as formed, the danger of tropicalization of Earth would recede. Instead of having the carbon dioxide level double, and Earth turn tropical, in 350 years, it would take 350×50 or 17,500 years to do so, and, heck, in that time, we'll think of something—we'll think of something——

However, the point of equilibrium is one thing and usually fairly easy to determine. The *rate at which equilibrium is reached* is quite another and often difficult to determine.

Yes, the ocean can dissolve the 6 billion tons of carbon dioxide we produce each year by burning coal and oil. There is plenty of room for it there. The ocean can hold 8 million times that quantity as a very minimum, over and above what it already holds. (This may create trouble for the fish, etc., but in eight million years we can solve that, perhaps.)

Nevertheless, though the oceans can dissolve it, will they do so quickly enough? If they will dissolve that quantity in a year, they keep pace with us, and all is well. If they dissolve it in a thousand years, we have produced 6,000 billion tons of carbon dioxide (probably much more) meanwhile, and we are out of luck.

But then why shouldn't the ocean dissolve the carbon dioxide quickly? The gas is soluble enough and there's water enough and to spare in the oceans. What's to stop it?

Ah, you see the solution only takes place at the surface of the ocean where air and water meet. If the surface skin gets loaded with carbon dioxide, no more will dissolve. It won't matter that the water just under the skin is empty. The rate of solution will then depend on how fast the carbon dioxide molecules drift downward out of the skin, or how fast the ocean water moves about so that fresh water reaches the skin where it can dissolve additional carbon dioxide.

The latter process seems to solve the problem, since we all know that the ocean is always in a lashing turmoil. Surely, then, it is well-mixed, with new water reaching the surface all the time.

Right—if we consider only the top 600 feet of ocean. Just as all the storms of our atmosphere are confined to the troposphere (the lowest 5 to 10 miles) so all the wild water movements of the ocean are confined to the top 600 feet or less. Below 600 feet there is only a slow, majestic movement, exactly how slow and majestic we are not yet certain. The rate of carbon dioxide solution, then, depends on how quickly that deep water (representing about 94 per cent of the total ocean volume) is brought to the surface.

There is some sort of circulation between the depths and the surface, we know. After all, the ocean can't dissolve oxygen by any means more magical than those by which it dissolves carbon dioxide, and yet we know there is oxygen dissolved in the ocean all the way down to the lowest abysses. We know, because there is animal life in those abysses that could not live in the absence of oxygen.

The longer the water stays down in the depths without renewal, the lower the oxygen concentration becomes through consumption by living organisms. This offers one method of following water circulation in the abyss. Bring up samples of deep water from, say, 3 miles down, and measure the oxygen content. The higher the oxygen content, the more recently that water was at the surface.

Such measurements have been made, and it turns out that the deep water with highest oxygen content is in the North Atlantic and around Antarctica. Apparently that is where surface water sinks to the bottom most readily. On the bottom there seems to be a slow movement that carries the water out of the Atlantic, around Africa, into the Indian Ocean, through the South Seas and into the Pacific—with the oxygen content declining constantly.

Granted that such an abyssal circulation does exist, how fast does it move? We might find out by adding something to the top of the ocean which is not already present in the ocean. Then we would have to wait for it to show up in various portions of the abyss and note the elapsed time. Of course, the added something would have to be detectable in extremely tiny quantities after we have allowed for dilution by an entire oceanful of water.

Actually, there may be something that fits the bill—strontium-90. There is a detectable quantity in the atmosphere now and there wasn't any fifteen years ago. Some has gotten into the ocean's skin but is there any already in the deep waters? If so, where? Chemists are devising methods of concentrating and measuring the strontium-90 in the ocean for just this purpose.

It would be odd if it turned out that the dangerous ash, strontium-90, were to give us vital information involving the dangers of the "harmless" ash, carbon dioxide. It's an ill wind——

(The abyssal circulation is important not only with respect to information concerning the carbon dioxide cycle. The lower waters are richer
in minerals—hence more fertile—than the life-scavenged upper waters.
If the time comes when man depends on the sea for most of his food, a
knowledge of abyssal circulation may be vital for "ocean-farming.")

Of course, why theorize as to how fast the ocean *may* dissolve carbon
dioxide, how slowly the atmospheric carbon dioxide *may* be building up,
how quickly the Earth *may* be turning into an iceless tropical world.
Why not actually measure the icecaps of the world and see if they are
disappearing or not. And if they are disappearing, how quickly? This, in
fact, was one of the prime objects of research for the Geophysical Year
and one of the more important reasons for all those scientists setting
up housekeeping on the Antarctic icecap.

We might also measure the actual over-all temperature of the Earth
and see if it is going up. If all the combusted carbon dioxide stays in
the atmosphere, while dissolving in the oceans at only a negligible rate,
then the over-all temperature ought to go up $1.1°$ C per century.

According to Gilbert N. Plass of Johns Hopkins, such temperature
measurements as are available indicate that just this rate of temperature
increase has indeed been going on since 1900. Of course, temperature
measurements during the first half of the twentieth century are not
reliable outside the more industrialized countries, so maybe this apparent increase matches the theoretical only through a coincidence aris-
ing from insufficient data.

However, if this is more than coincidence; if Earth is really warming
up at that rate, then wave good-by to the icecaps. And if you live at the
seashore, your not too distant descendants may well have to visit the
old homestead with a skin-diver's outfit.

Earth has survived a similiar fate three times in the last 300,000
years, this current rise being the fourth. These periods of rise make up
what are called the "interglacial epochs." Earth has also survived four
periods of temperature drop in this same period of time, each of which
initiated a "glacial epoch" or, as it is more commonly known, an "ice
age." It might seem that there is some physical phenomenon which
brings on this coming and going of the ice, and one would expect that
same phenomenon to continue and to keep the oscillation of ice and
no-ice going for the immediate future (by which I mean the next few
million years).

Yet prior to 300,000 years ago (for at least 200,000,000 years prior, in fact) there were no Ice Ages. For all that long period (or more) Earth was reasonably ice-free. Naturally, the question arises: what happened 300,000 years ago?

One explanation is that Earth undergoes a temperature oscillation of a very slow and majestic type which didn't make itself visible in the form of ice till 300,000 years ago. For instance, a Serbian physicist named Milutin Milankovich in the 1920s suggested that because of oscillations in Earth's orbit and the tilt of its axis, the planet picks up a bit more heat from the Sun at some times than at others. His proposed temperature cycle lasted 40,000 years, so that there is a kind of 20,000-year long "Great Summer" and a 20,000-year long "Great Winter." The temperature differences involved are not really very great but, as I stated earlier a drop of less than 4° C in Earth's present temperature would be enough to kick off an Ice Age.

This Milankovich-oscillation can be made to explain the recent advances and retreats of the glaciers, but what about the situation B.I.A. (Before the Ice Ages)?

Well, what if prior to 300,000 years ago, Earth's over-all temperature were sufficiently high so that even the Great Winter dip was not enough to bring on the ice? You can see that, if you consider the annual temperature oscillation between ordinary summer and winter. In New York this oscillation crosses the freezing point of water, so there is rain in the summer but snow in the winter. In Miami the average temperature is higher and the oscillation does not dip low enough to bring snow in the winter. On a planetary scale, what if Earth's climate switched from ice-less Miami to periodically icy New York?

This possibility has been checked by isotope analysis. (These days, if a scientist can't get an answer by isotope analysis, he ain't hep.) There are three stable oxygen isotopes: oxygen-16, which makes up 99.76 per cent of all the oxygen atoms; oxygen-18 (0.20 per cent) and oxygen-17 (0.04 per cent). They all behave almost alike, so alike that ordinarily no difference can be detected. However, oxygen-18 is 12½ per cent heavier than oxygen-16 and correspondingly slower in its reactions. For instance, when water evaporates, water molecules containing oxygen-16 get into the air a bit more easily than those containing oxygen-18, and if evaporation continues over a long interval the water that is left contains noticeably more oxygen-18 than it had originally.

This applies to the oceans, which are constantly evaporating, so that sea water should (and does) have a bit more oxygen-18, in proportion to oxygen-16, than does fresh water, which is made up of the evaporated portion of the oceans. Furthermore, this effect is increased as the temperature goes up. For each 1° C rise in the temperature of the ocean, the ratio of oxygen-18 to oxygen-16 goes up 0.02 per cent.

Now then, fossil sea shells are made up largely of calcium carbonate. The calcium carbonate contains oxygen atoms which were derived from the ocean water. The oxygen-18/oxygen-16 ratio in those shells must therefore reflect the ratio in the water from which they derived the oxygen and that, in turn, should give a measure of the ocean temperatures of ages long past.

Such measurements were first made in the laboratories of Harold C. Urey at the University of Chicago and proved a very tricky job. On the basis of such measurements, however, it turns out that during the Mesozoic Age of old, when dinosaurs were bold, the ocean temperatures were as high as 21° C (70° F).

This bespeaks a planetary temperature too high to allow·an Ice Age, even at the bottom of the Milankovich cycle.

But then, beginning 80,000,000 years ago, when ocean temperatures were at the 21° peak, the temperatures started dropping and have continued to do so ever since.

According to Cesare Emiliani (who carried temperature measure-

HAROLD CLAYTON UREY

Urey was born in Walkerton, Indiana, on April 29, 1893. At the University of Montana he majored in zoology, graduating in 1917. Work during World War I, however, turned his attention to high explosives and through that to chemistry generally.

In 1931 Urey tackled the problem of heavy hydrogen. There had been suggestions that there might be a form of hydrogen with atoms twice the mass of the ordinary hydrogen atom, but if it existed at all, it had to be present only in minute concentrations.

It seemed to Urey that heavy hydrogen ought to boil less readily than ordinary hydrogen. He therefore began with four liters of liquid hydrogen and then let it boil very slowly until only a single cubic centimeter remained. This last dreg ought to be much richer in heavy hydrogen, if that existed, than the original was. He tested it spectroscopically and sure enough hydrogen lines, displaced as they ought to be (according to theory) for the heavier atoms, showed up. Urey was awarded the 1934 Nobel Prize in chemistry for this feat.

Urey then began to investigate methods of separating isotopes of other elements and made use of the fact that heavier isotopes tended to react a bit more slowly than their lighter twins. He was able to prepare samples of carbon and nitrogen which were enriched in the heavier isotopes. This isotope-separation technique proved useful during World War II when the atom bomb was being developed.

After the war, Urey turned away deliberately from work that could be turned to destructive use and became interested in the geochemistry of the Earth and the Moon. Here, too, isotopes proved useful. The proportion of oxygen isotopes in seashells depends on temperature and by studying fossil shells, the temperature of the ancient oceans could be determined.

For what it's worth, Urey tried to teach me thermodynamics back in 1940 (with, I think, only limited success).

ments into the recent past, geologically speaking) the explanation for this is that, after a long period of land area fairly free of mountains and oceans fairly free of abyss, so that many shallow seas covered much of what is now land, a geological revolution occurred. The ocean bottoms started sinking and mountain ranges started rising.

With land going up and ocean bottoms down, new land was exposed very gradually. Land stores less heat than does water, radiating more away at night, so that the Earth's over-all temperature gradually dropped. Also, new land meant new rocks exposed to carbon dioxide weathering, which meant a fall in the carbon dioxide of the atmosphere, a decrease of the "greenhouse effect," and again, a fall in temperature.

Quite possibly it was this fall in temperature that killed off the dinosaurs.

By a million years ago, the steady drop of ocean temperatures had brought it down to $2°$ C ($35\frac{1}{2}°$ F) and by 300,000 years ago, Earth's temperature was low enough to allow the Ice Ages to appear at the bottom of the Milankovich cycles.

A somewhat more startling explanation of the beginning of the Ice Ages has been advanced by Maurice Ewing and William Donn, working at Columbia. They blame it specifically on the Arctic Ocean.

The North Pole is located in a small, nearly landlocked arm of the ocean, which is small enough and landlocked enough to make possible an unusual state of affairs.

Thus the suggestion is that when the Arctic Ocean is free of ice, it acts as a reservoir of evaporating water that feeds snow storms in the winter. If the Arctic Ocean were large and open, most of these snow storms would fall on the open sea and there melt. As it is, the snow falls upon the surrounding land areas of Canada and Siberia, and because of the lower heat content of land areas, it does not melt but remains during the winter. In fact, it accumulates from winter to winter, with the summer never quite melting all the ice produced by the preceding winter. The glaciers form and creep southward.

Once this happens, a considerable fraction of the Earth is covered with ice, which reflects more of the Sun's radiation than does either land or water. Furthermore, the Earth as a whole is cloudier and stormier during an Ice Age than otherwise, and the excess clouds also reflect more of the Sun's radiation. Altogether, about 7 per cent of the Sun's radiation, that would ordinarily reach Earth, is reflected during an Ice Age. The Earth's temperature drops and the Arctic Ocean, which (according to Ewing and Donn) remained open during the height of glacier activity, finally freezes over. (Even despite the lowering temperature, it does this only because it is so small and landlocked.)

Once the Arctic freezes over, the amount of evaporation from it is drastically decreased, the snowstorms over Canada and Siberia are cut down, the summers (cool as they are) suffice to melt more than the decreased accumulation, and the glaciers start retreating. The Earth warms up again (as it is now doing), the Arctic Ocean melts (this point not yet having been reached in the current turn of the cycle), the snows begin again, and bang comes another glaciation.

But why did all this only start 300,000 years ago? Ewing and Donn say because that is when the North Pole first found itself in the Arctic Ocean. Before then it had been somewhere in the Pacific where the ocean was large enough and open enough to cause no severe snowstorms on the distant land areas.

Ice Ages could continue to annoy us periodically, then, until the present mountains wear down to nubs and the ocean bottoms rise, or until the North Pole leaves the Arctic (depending on which theory—if either—is correct).

Unless, that is, something new interferes, such as the carbon dioxide we are pouring into the atmosphere. The current temperature rise is being radically hastened, apparently, by the increased carbon dioxide in the atmosphere. The next temperature drop may be correspondingly slowed and may, conceivably, not drop far enough to start a new glaciation.

Therefore, it is possible that Earth has seen its last Ice Age, regardless of the Milankovich cycle or the position of the North Pole, until such time as the ocean, or we ourselves, can get rid of the excess carbon dioxide once again. Within a matter of centuries, then, we may reverse much or all of an 80,000,000 year trend of dropping temperature and find ourselves back in the Mesozoic, climatically speaking, only without (thank goodness) the dinosaurs.[1]

[1] *This article was written in 1959. It did not take into account that industrialization is belching dust of all kinds into the air as well as carbon dioxide. The dust blocks some of the sunlight reaching Earth and exerts a cooling effect. So far the dust-cooling is overreaching the carbon-dioxide-heating and world temperatures have actually been dropping since 1940. If we ever clean up the air, however, watch out! The temperature may bound upward and the icecaps may start melting with dismaying rapidity.*

PART VI

GENERAL

16

THE NOBELMEN OF SCIENCE

Something happened to me some time ago which still leaves me stunned.

I got a call early in the morning from a reporter. He said, "Three Frenchmen have just won the Nobel Prize for Medicine and Physiology for their work on genetics, and I thought you might explain to me, in simple terms, the significance of their discoveries."

"Who are the three Frenchmen?" I asked.

He told me, and the names drew a blank. I pleaded ignorance of their work and apologized. He hung up.

I sat there for a while and brooded, since I hate revealing flaws in my omniscience. As is usual in such cases, I thought long, shivery thoughts about the oncoming of senility, and then cudgeled my brain unmercifully. What discoveries in genetics in recent years, I wondered, would rate a Nobel Prize this year?

One thought came to mind. Some years back it had been discovered that mongolism (a kind of congenital mental retardation marked by a variety of characteristic symptoms) was accompanied by an additional chromosome in the cells. The twenty-first pair had three members rather than two, giving a total number of forty-seven rather than forty-six.

I ripped into my library to find the record of the discovery and came upon it in a matter of minutes. Eureka! It had first been reported by three Frenchmen!

My chest expanded; my cheeks glowed; my brain palpitated. There was life, it seemed, in the old boy yet.

I called the reporter back, savoring my triumph in advance. "What were the names of those Frenchmen, again?" I asked cheerfully.

He rattled them off. *They were three different names!* There was a long dismal pause, and then I said, "'Sorry. I still don't know."

He must be wondering, ever since, why I bothered calling him back. As for me, I faced the situation staunchly; I got back into bed and pulled the covers over my head.

(If you're curious, the three Frenchmen who won the 1965 prize—who will be named later in the article—received it for the discovery that some genes have regulatory functions and control the activity of other genes.)

It got me to thinking about Nobel Prize winners, however, and the way such victories have become a matter of national pride. After all, the reporter didn't identify the winners as "three geneticists" but as "three Frenchmen."

But in what way ought nations to take credit? Einstein, for instance, was born in Germany, but was educated in Italy and Switzerland as well as in Germany, was in Germany at the time he won the award, though he was in Switzerland at the time he wrote his first world-shattering papers, and he spent his later life as an American citizen. How do we list him?

I would like to suggest that the key point is a man's scientific birth and that this takes place in college. Schooling before college is too vague and diffuse, schooling after college involves a man who is already set in his directions. It is college itself that sets those directions and determines whether a man will throw himself into science or not.

Naturally, I am aware that influences prior to college and outside college may be very important but here we edge uncomfortably close to the task of making a psychiatric study of each Nobel Prize winner, which I can't and won't do. I am going to accept college as a "first approximation" and list the winners by the nationality of that college.

This isn't easy, either.

In the first place, the information I can dig up isn't always clear in this matter and I may make some mistakes. Secondly, some Nobelists went to two different undergraduate colleges in two different nations and I have to choose among them. Thirdly, deciding on a nationality brings me face to face with the fact that political divisions change their nature with time. Some colleges that were once located in Austria-Hungary are now located in Czechoslovakia, without having ever budged.

One Nobel Prize winner studied at the University of Dorpat, and this is perhaps the most troublesome case. The university was in Russia then and it is in the Soviet Union now. In between, however, it was in Estonia, which the American government still recognizes as an independent nation. And to top it off, I'm sure that both the winner himself and the university, in his time, were completely German in a cultural sense.

Oh, well, I will take full responsibility for my decisions on categories and I am certain that those Gentle Readers who take issue with those decisions will write me and tell me so. And I may be argued into making changes.

So now let's make the lists which, if they make dull reading, have the advantage of not being like any other list in existence (as far as I know) and which therefore come under the heading, I hope, of "valuable reference."

We'll start with the Nobel Prize in physics in Table 27:

Table 27—The Physics Nobelists

YEAR	WINNER(S)—PHYSICS	UNDERGRADUATE TRAINING
1901	Wilhelm Konrad Röntgen	Switzerland
1902	Hendrik Antoon Lorentz	Netherlands
	Pieter Zeeman	Netherlands
1903	Antoine Henri Becquerel	France
	Pierre Curie	France
	Marie Sklodowska Curie	France
1904	John William Strutt, Baron Rayleigh	Great Britain
1905	Philipp Lenard	Austria

MARIE SKLODOWSKA CURIE

Marie Sklodowska was born in Warsaw, Poland, on November 7, 1867. Both her parents were highly educated, but Poland was under Russian domination at the time and a Polish revolt in 1863 had made the Russian fist clench harder. To obtain education for herself, it was necessary for her to leave Poland. In 1891 she went to Paris where she entered the Sorbonne to study physical chemistry, living with the greatest frugality (so that she fainted with hunger in the classroom at one time) but graduating at the top of the class. On July 26, 1895, she married Pierre Curie, a French physicist who had already made important discoveries.

Marie Curie began the investigation of the new field of radiations produced by uranium (she coined the word "radioactivity") and showed that it was indeed the uranium atom that was the source of the radiations. She also found that the element thorium was radioactive. Pierre Curie, recognizing the talent of his wife, abandoned his own researches and joined her in her work.

The Curies noted that uranium ores were more radioactive than could be accounted for by the uranium atoms alone and began a search for other and more radioactive elements that might be present there. By July of 1898, the two had isolated from uranium ore a small pinch of powder containing a new element hundreds of times as radioactive as uranium. This was called "polonium" after Marie's native land. In December 1898, they detected the still more radioactive "radium." They spent four years isolating a gram of radium salts from tons of uranium ore. In 1903 Marie obtained her Ph.D. with a dissertation describing this work, and promptly she and Pierre won a share in the Nobel Prize in physics for it.

In 1906 Pierre was killed in a traffic accident, being run over by a horse-drawn vehicle, and Marie took over his professorship at the Sorbonne—the first woman ever to hold such a post. In 1911 she was awarded the Nobel Prize for chemistry for her discovery of polonium and radium. She died of leukemia in Haute Savoie, France, on July 4, 1934.

YEAR	WINNER(S)—PHYSICS	UNDERGRADUATE TRAINING
1906	Joseph John Thomson	Great Britain
1907	Albert Abraham Michelson	United States
1908	Gabriel Lippmann	France
1909	Guglielmo Marconi	Italy (private tutoring)
	Karl Ferdinand Braun	Germany
1910	Johannes Diderik van der Waals	Netherlands
1911	Wilhelm Wien	Germany
1912	Nils Gustaf Dalén	Sweden
1913	Heike Kamerlingh Onnes	Netherlands
1914	Max Theodor Felix von Laue	Germany
1915	William Henry Bragg	Great Britain
	William Lawrence Bragg	Australia
1916	(no award)	
1917	Charles Glover Barkla	Great Britain
1918	Max Karl Ernst Ludwig Planck	Germany
1919	Johannes Stark	Germany
1920	Charles Edouard Guillaume	Switzerland
1921	Albert Einstein	Switzerland
1922	Niels Henrik David Bohr	Denmark
1923	Robert Andrews Millikan	United States
1924	Karl Manne Georg Siegbahn	Sweden
1925	James Franck	Germany
	Gustav Hertz	Germany
1926	Jean Baptiste Perrin	France
1927	Arthur Holly Compton	United States
	Charles Thomson Rees Wilson	Great Britain

YEAR	WINNER(S)—PHYSICS	UNDERGRADUATE TRAINING
1928	Owen Willans Richardson	Great Britain
1929	Prince Louis-Victor de Broglie	France
1930	Chandrasekhara Venkata Raman	India
1931	(no award)	
1932	Werner Karl Heisenberg	Germany
1933	Erwin Schrödinger Paul Adrien Maurice Dirac	Austria Great Britain
1934	(no award)	
1935	James Chadwick	Great Britain
1936	Victor Francis Hess Carl David Anderson	Austria United States
1937	Clinton Joseph Davisson George Paget Thomson	United States Great Britain
1938	Enrico Fermi	Italy
1939	Ernest Orlando Lawrence	United States
1940	(no award)	
1941	(no award)	
1942	(no award)	
1943	Otto Stern	Germany
1944	Isidor Isaac Rabi	United States
1945	Wolfgang Pauli	Austria
1946	Percy Williams Bridgman	United States
1947	Edward Victor Appleton	Great Britain
1948	Patrick Maynard Stuart Blackett	Great Britain
1949	Hideki Yukawa	Japan
1950	Cecil Frank Powell	Great Britain

YEAR	WINNER(S)—PHYSICS	UNDERGRADUATE TRAINING
1951	John Douglas Cockcroft	Great Britain
	Ernest Thomas Sinton Walton	Ireland
1952	Felix Bloch	Switzerland
	Edward Mills Purcell	United States
1953	Fritz Zernike	Netherlands
1954	Max Born	Germany
	Walter Bothe	Germany
1955	Willis Eugene Lamb, Jr.	United States
	Polykarp Kusch	United States
1956	William Bradford Shockley	United States
	John Bardeen	United States
	Walter Houser Brattain	United States
1957	Tsung Dao Lee	China
	Chen Ning Yang	China
1958	Pavel Alekseyevich Cherenkov	Soviet Union
	Ilya Mihailovich Frank	Soviet Union
	Igor Yevgenyevich Tamm	Soviet Union
1959	Emilio Segrè	Italy
	Owen Chamberlain	United States
1960	Donald Arthur Glaser	United States
1961	Robert Hofstadter	United States
	Rudolf Ludwig Mössbauer	Germany
1962	Lev Davidovich Landau	Soviet Union
1963	Eugene Wigner	Germany
	J. Hans Daniel Jensen	Germany
	Maria Goeppert-Mayer	Germany
1964	Charles Hard Townes	United States
	Nikolai Basov	Soviet Union
	Alexander Prochorov	Soviet Union
1965	Julian Seymour Schwinger	United States
	Richard Phillips Feynman	United States
	Shinichiro Tomonaga	Japan

Next we'll take up the Nobel Prizes in chemistry in Table 28:

Table 28—The Chemistry Nobelists

YEAR	WINNER(S)—CHEMISTRY	UNDERGRADUATE TRAINING
1901	Jacobus Henricus Van't Hoff	Netherlands
1902	Emil Fischer	Germany
1903	Svante August Arrhenius	Sweden
1904	William Ramsay	Great Britain
1905	Adolf von Baeyer	Germany
1906	Henri Moissan	France
1907	Eduard Buchner	Germany
1908	Ernest Rutherford	New Zealand
1909	Wilhelm Ostwald	Russia
1910	Otto Wallach	Germany
1911	Marie Sklodowska Curie	France
1912	Victor Grignard	France
	Paul Sabatier	France
1913	Alfred Werner	Switzerland
1914	Theodore William Richards	United States
1915	Richard Willstätter	Germany
1916	(no award)	
1917	(no award)	
1918	Fritz Haber	Germany
1919	(no award)	
1920	Walther Hermann Nernst	Germany
1921	Frederick Soddy	Great Britain
1922	Francis William Aston	Great Britain
1923	Fritz Pregl	Austria

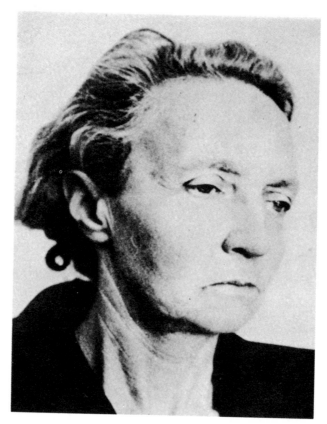

IRÈNE JOLIOT-CURIE
FRÉDÉRIC JOLIOT-CURIE

Irène Curie was the elder daughter of Pierre and Marie Curie, and was born in Paris on September 12, 1897, as her parents were beginning their monumental researches on radioactivity. While working as her mother's assistant, she met Frédéric Joliot, another assistant (born in Paris, March 19, 1900) and they were married in 1926. Since the Curies had no sons and since Joliot did not wish that eminent name to die out, he changed his name to Joliot-Curie.

When ill health forced her mother to retire, Irène succeeded to her post. She and Frédéric worked together (as her parents had done) on further researches into radioactivity.

In 1934 they were studying the effect of alpha particles on light elements such as aluminum. They knocked protons out of the aluminum nuclei as was to be expected. At one point, however, the Joliot-Curies discovered that after they had ceased alpha particle

bombardment and protons ceased being emitted, another form of radiation continued.

Close study showed that in bombarding aluminum they had formed phosphorus, but not the ordinary form that existed in nature. They had formed an artificial radioactive isotope of phosphorus. For this discovery of "artificial radioactivity" (and over a thousand radioactive isotopes of the various elements have since been formed), they received the Nobel Prize in chemistry in 1935.

Irène served a short period in the cabinet of Léon Blum in 1936 and both she and Frédéric were politically active leftists. After the German conquest of France in 1940, the Joliot-Curies managed to smuggle the heavy water necessary for atomic bomb research out of the country, but they themselves remained behind in order to help organize the Resistance to Hitler. Irène finally left for Switzerland in 1944.

Both died in Paris, Irène on March 17, 1956 (of leukemia, like her mother, probably induced by overexposure to energetic radiation) and Frédéric on August 14, 1958.

YEAR	WINNER(S)—CHEMISTRY	UNDERGRADUATE TRAINING
1924	(no award)	
1925	Richard Zsigmondy	Austria
1926	Theodor Svedberg	Sweden
1927	Heinrich Wieland	Germany
1928	Adolf Windaus	Germany
1929	Arthur Harden Hans von Euler-Chelpin	Great Britain Germany
1930	Hans Fischer	Germany
1931	Carl Bosch Friedrich Bergius	Germany Germany
1932	Irving Langmuir	United States
1933	(no award)	
1934	Harold Clayton Urey	United States
1935	Irène Joliot-Curie Frédéric Joliot-Curie	France France
1936	Peter Joseph Wilhelm Debye	Germany
1937	Walter Norman Haworth Paul Karrer	Great Britain Switzerland
1938	Richard Kuhn	Germany
1939	Adolf Butenandt Leopold Ruzicka	Germany Switzerland
1940	(no award)	
1941	(no award)	
1942	(no award)	
1943	Georg von Hevesy	Hungary
1944	Otto Hahn	Germany
1945	Artturi Ilmari Virtanen	Finland

YEAR	WINNER(S)—CHEMISTRY	UNDERGRADUATE TRAINING
1946	James Batcheller Sumner John Howard Northrop Wendell Meredith Stanley	United States United States United States
1947	Robert Robinson	Great Britain
1948	Arne Tiselius	Sweden
1949	William Francis Giauque	United States
1950	Otto Diels Kurt Alder	Germany Germany
1951	Glenn Theodore Seaborg Edwin Mattison McMillan	United States United States
1952	Archer John Porter Martin Richard Laurence Millington Synge	Great Britain Great Britain
1953	Hermann Staudinger	Germany
1954	Linus Carl Pauling	United States
1955	Vincent Du Vigneaud	United States
1956	Cyril Norman Hinshelwood Nikolai Nikolaevitch Semënov	Great Britain Soviet Union
1957	Alexander Robertus Todd	Great Britain
1958	Frederick Sanger	Great Britain
1959	Jaroslav Heyrovsky	Czechoslovakia
1960	Willard Frank Libby	United States
1961	Melvin Calvin	United States
1962	Max Ferdinand Perutz John Cowdery Kendrew	Austria Great Britain
1963	Karl Ziegler Giulio Natta	Germany Italy
1964	Dorothy Crowfoot Hodgkin	Great Britain
1965	Robert Burns Woodward	United States

GEORG VON HEVESY

Georg von Hevesy was born in Budapest, Hungary, on August 1, 1885, and is notable for two discoveries in chemistry, both of which can be pinned down to the year 1923.

Hevesy was working at a time when most of the stable elements had been discovered and little room was left for the dramatic announcements of element after element that had marked the nineteenth century. Working in Copenhagen under Niels Bohr, however, Hevesy followed Bohr's suggestion that undiscovered element No. 72

Finally, the Nobel Prizes for medicine and physiology in Table 29:

Table 29—The Medicine and Physiology Nobelists

YEAR	WINNER(S)—MED. AND PHYSIOL.	UNDERGRADUATE TRAINING
1901	Emil von Behring	Germany
1902	Ronald Ross	Great Britain
1903	Niels Ryberg Finsen	Denmark
1904	Ivan Petrovich Pavlov	Russia
1905	Robert Koch	Germany
1906	Camillo Golgi	Italy
	Santiago Ramón y Cajal	Spain
1907	Charles Louis Alphonse Laveran	France
1908	Ilya Mechnikov	Russia
	Paul Ehrlich	Germany

should be very similar to zirconium in chemical properties and ought therefore be sought in zirconium ores. He carried through this investigation and in January 1923 announced the discovery of "hafnium," a name derived from the Latinized version of Copenhagen.

Hevesy was also interested in using radioactive atoms to study living systems. A radioactive isotope could not be distinguished from the stable isotopes of the same element by ordinary chemical reactions (including those that take place in living tissue) but was easily detected by devices sensitive to subatomic particles. In 1918 he used radioactive lead which he incorporated into compounds and dissolved in water in small quantities. By watering plants with solutions containing the radioactive isotope, he was able in 1923 to follow the absorption and distribution of lead under these circumstances in great detail.

Radioactive lead was not in itself very useful as a tool for studying living systems, but it lead the way to the use of radioactive isotopes generally, a technique that was to revolutionize biochemistry.

In 1943 Hevesy fled Hitler's armies and found refuge in neutral Sweden. While there he was awarded the 1943 Nobel Prize in chemistry.

He died in Freiburg, Germany, on July 5, 1966.

YEAR	WINNER(S)—MED. AND PHYSIOL.	UNDERGRADUATE TRAINING
1909	Emil Theodor Kocher	Switzerland
1910	Albrecht Kossel	Germany
1911	Allvar Gullstrand	Sweden
1912	Alexis Carrel	France
1913	Charles Richet	France
1914	Robert Bárány	Austria
1915	(no award)	
1916	(no award)	
1917	(no award)	
1918	(no award)	
1919	Jules Bordet	Belgium
1920	August Krogh	Denmark
1921	(no award)	
1922	Archibald Vivian Hill Otto Meyerhof	Great Britain Germany
1923	Frederick Grant Banting John James Richard Macleod	Canada Great Britain
1924	Willem Einthoven	Netherlands
1925	(no award)	
1926	Johannes Fibiger	Denmark
1927	Julius Wagner-Jauregg	Austria
1928	Charles Nicolle	France
1929	Christiaan Eijkman Frederick Gowland Hopkins	Netherlands Great Britain
1930	Karl Landsteiner	Austria
1931	Otto Warburg	Germany
1932	Charles Scott Sherrington Edgar Douglas Adrian	Great Britain Great Britain

YEAR	WINNER(S)—MED. AND PHYSIOL.	UNDERGRADUATE TRAINING
1933	Thomas Hunt Morgan	United States
1934	George Hoyt Whipple George Richards Minot William Parry Murphy	United States United States United States
1935	Hans Spemann	Germany
1936	Henry Hallett Dale Otto Loewi	Great Britain Germany
1937	Albert von Szent-Györgyi	Hungary
1938	Corneille Heymans	Belgium
1939	Gerhard Domagk	Germany
1940	(no award)	
1941	(no award)	
1942	(no award)	
1943	Henrik Dam Edward Adelbert Doisy	Denmark United States
1944	Joseph Erlanger Herbert Spencer Gasser	United States United States
1945	Alexander Fleming Ernst Boris Chain Howard Walter Florey	Great Britain Germany Australia
1946	Hermann Joseph Muller	United States
1947	Bernardo Alberto Houssay Carl Ferdinand Cori Gerty Theresa Radnitz Cori	Argentina Czechoslovakia Czechoslovakia
1948	Paul Müller	Switzerland
1949	Walter Rudolf Hess Antonio Egas Moniz	Switzerland Portugal
1950	Edward Calvin Kendall Philip Showalter Hench Tadeus Reichstein	United States United States Switzerland

YEAR	WINNER(S)—PHYSICS AND PHYSIOL.	UNDERGRADUATE TRAINING
1951	Max Theiler	South Africa
1952	Selman Abraham Waksman	United States
1953	Fritz Albert Lipmann	Germany
	Hans Adolf Krebs	Germany
1954	John Franklin Enders	United States
	Thomas Huckle Weller	United States
	Frederick Chapman Robbins	United States
1955	Axel Hugo Teodor Theorell	Sweden
1956	Dickinson Woodruff Richards, Jr.	United States
	André Frédéric Cournand	France
	Werner Theodor Otto Forssmann	Germany
1957	Daniel Bovet	Switzerland
1958	George Wells Beadle	United States
	Edward Lawrie Tatum	United States
	Joshua Lederberg	United States
1959	Severo Ochoa	Spain
	Arthur Kornberg	United States
1960	Macfarlane Burnet	Australia
	Peter Brian Medawar	Great Britain
1961	Georg von Bekesy	Switzerland
1962	Francis Harry Compton Crick	Great Britain
	Maurice Hugh Frederick Wilkins	Great Britain
	James Dewey Watson	United States
1963	John Carew Eccles	Australia
	Alan Lloyd Hodgkin	Great Britain
	Andrew Fielding Huxley	Great Britain
1964	Konrad Bloch	Germany
	Feodor Lynen	Germany
1965	André Lwoff	France
	Jacques Monod	France
	François Jacob	France

Now for some overall statistics—59 physics prizes have been shared among 86 people (including 2 women); 57 chemistry prizes have been shared among 71 people (including 3 women); 56 medicine and physiology prizes have been shared among 90 people (including 1 woman). All told, 246 people (including 5 women) have won among them a total of 172 prizes in the sciences through 1965. (If this sum seems wrong to you, I must remind you that Marie Sklodowska Curie won two science prizes, one in physics and one in chemistry.) Suppose we next list the prizes according to the college nationality of

Table 30—Nobelists by Nationality

NATION	PHYSICS	CHEMISTRY	MEDICINE & PHYSIOLOGY	TOTAL
Germany	10	17½	10⅙	37⅔
United States	12	11	9⅙	32⅙
Great Britain	10½	10	6⅙	26⅔
France	4	4	5½	13½
Switzerland	3½	2	4⅚	10⅓
Austria	3	2½	3	8½
Sweden	2	3	2	7
Netherlands	4	1	1½	6½
Soviet Union (Russia)	2⅔	1½	1½	5⅔
Denmark	1	—	3½	4½
Italy	2	½	½	3
Hungary	—	1	1	2
Belgium	—	—	2	2
Australia	½	—	1⅙	1⅔
Czechoslovakia	—	1	⅔	1⅔
Japan	1⅓	—	—	1⅓
India	1	—	—	1
China	1	—	—	1
New Zealand	—	1	—	1
Finland	—	1	—	1
Spain	—	—	1	1
South Africa	—	—	1	1
Ireland	½	—	—	½
Portugal	—	—	½	½
Canada	—	—	½	½
Argentina	—	—	⅓	⅓
Totals	59	57	56	172

the winners. In making this final table, I will give half-credit to each man who shares the prize with one other and third-credit to each man who shares the prize with two others. Sometimes the cash award of the prize is divided $\frac{1}{2}$, $\frac{1}{4}$, and $\frac{1}{4}$, but I'll ignore that refinement.

The result is in Table 30.

Too much must not be read into this table. It can't and should not be used to indicate anything about the relative intelligences of the nations. There is a wide difference in quality between one Nobelist and another and these differences are impossible to weigh objectively. For instance, Italy has a score of only 3 altogether, but one of those three was scored by Enrico Fermi. Denmark only has 1 physics prize to its credit, but that one was Niels Bohr, a giant even among giants.

What the table does show is that the scientific tradition has been entrenched longest and most strongly in the educational systems of those nations culturally related to Great Britain and Germany. As nearly as I can estimate, colleges that teach in either English, German, Swedish, Dutch, or Danish account for $135\frac{2}{3}$, or 80 per cent, of the Nobel Prizes in the sciences.

I suspect that this English-German domination of science will grow progressively less marked in succeeding decades.

And I would like to make one final pitch, too. It is about time that the arbitrary decision of Nobel that only three classifications of science deserve the prize be reconsidered. I'm thinking of three more classifications: (a) Astronomy (to include geology and oceanography), (b) Mathematics, and (c) Science of man (to include anthropology, psychology, and, possibly, sociology).

Many great scientists go unhonored by the present system and it bothers me. (And it probably bothers them, too.)[1]

[1] In 1969, three years after this article was written the Nobel Prizes were expanded with the addition of a prize for economics. This does not in the least diminish the need for the categories I mentioned.

ADDENDUM

Of all the articles in this volume, "The Nobelmen of Science" is most clearly dated. Written in 1966, it includes only the winners through 1965. It is obviously necessary to bring the list up to date through 1972. Therefore:

Table 31—The Physics Nobelists After 1965

YEAR	WINNER(S)—PHYSICS	UNDERGRADUATE TRAINING
1966	Alfred Kastler	France
1967	Hans Albrecht Bethe	Germany
1968	Luis W. Alvarez	United States
1969	Murray Gell-Mann	United States
1970	Hannes Alfven Louis Neel	Sweden France
1971	Dennis Gabor	Hungary
1972	John Bardeen Leon N. Cooper John Robert Schrieffer	United States United States United States

Table 32—The Chemistry Nobelists After 1965

YEAR	WINNER(S)—CHEMISTRY	UNDERGRADUATE TRAINING
1966	Robert S. Mullikan	United States
1967	Manfred Eigen Ronald G. W. Norrish George Porter	Germany Great Britain Great Britain
1968	Lars Onsager	Norway
1969	Derek H. R. Barton Odd Hassel	Great Britain Norway

YEAR	WINNER(S)—CHEMISTRY	UNDERGRADUATE TRAINING
1970	Luis Frederico Leloir	Argentina
1971	Gerhard Herzberg	Germany
1972	Christian B. Anfinsen	United States
	Stanford Moore	United States
	William H. Stein	United States

Table 33—The Medicine and Physiology Nobelists After 1965

YEAR	WINNER(S)—MEDICINE AND PHYSIOLOGY	UNDERGRADUATE TRAINING
1966	Francis Peyton Rous	United States
	Charles B. Huggins	Canada
1967	Haldan Keffer Hartline	United States
	George Wald	United States
	Ragnar A. Granit	Finland
1968	Robert W. Holley	United States
	Marshall W. Nirenberg	United States
	Har Gobind Khorana	India
1969	Max Delbrück	Germany
	Alfred D. Hershey	United States
	Salvador Luria	Italy
1970	Julius Axelrod	United States
	Bernard Katz	Germany
	Ulf Svante von Euler	Sweden
1971	Earl W. Sutherland, Jr.	United States
1972	Gerald M. Edelman	United States
	Rodney R. Porter	Great Britain

Now for some post-1965 statistics. Seven Nobel Prizes in each category have been shared among 10 people in physics; 12 people in chemistry; and 17 people in medicine and physiology. All 39 individuals were men; no women.

There have been 193 prizes awarded so far since 1901 and these have been distributed among 284 people altogether (including 5 women). I mentioned in the article that Marie Curie had won two science prizes, one in physics and one in chemistry. Now John Bardeen has done it, too, and become a double laureate, his prizes both being in physics.

Next I will prepare a table of Nobelists by nationality for the period after 1965, analogous to Table 30. This you will find in Table 34:

Table 34—Nobelists Since 1965 by Nationality

NATION	PHYSICS	CHEMISTRY	MEDICINE & PHYSIOLOGY	TOTAL
United States	3	2	4	9
Germany	1	1⅓	⅔	3
Great Britain	—	1⅙	½	1⅔
Norway	—	1½	—	1½
France	1½	—	—	1½
Hungary	1	—	—	1
Argentina	—	1	—	1
Sweden	½	—	⅓	⅚
Canada	—	—	½	½
Finland	—	—	⅓	⅓
India	—	—	⅓	⅓
Italy	—	—	⅓	⅓
Totals	7	7	7	21

As you see, since 1965 the United States has dominated the Nobel scene, particularly in medicine and physiology. Germany has been in second place, but if the total is taken from 1901 to 1972, combining Tables 30 and 34, the United States now has 41⅙ Prizes altogether, compared with Germany 40⅔ and it moves into all-round first place. Great Britain remains in third place.

No new nationalities are represented in the list since 1965 and if we consider colleges that teach in either English, German, Swedish, Dutch, or Danish (and throw in Norway and Finland to include all the Scandinavian nations), they still make up 76 per cent of the whole. Science is still, apparently, a specialty of the Teutonic nations, with France the most important non-Teutonic nation in the field. My suggestion in 1965 that this Teutonic domination would fade out may yet come to pass but there is no sign of that as yet.

17

THE ISAAC WINNERS

When one looks back over the months or years, it becomes awfully tempting to try to pick out the best in this or that category. Even the ancient Greeks did it, choosing the "seven wise men" and the "seven wonders of the world."

We ourselves are constantly choosing the ten best-dressed women of the year or the ten most notable news-breaks, or we list the American Presidents in order of excellence. The FBI and other law-enforcement agencies even list criminals in the order of their desirability (behind bars, that is).

There is a certain sense of power in making such lists. An otherwise undistinguished person suddenly finds himself able to make decisions with regard to outstanding people, taking this one into the fold and hurling that one into the outer darkness. One can, after some thought, move x up the list and y down, possibly changing the people so moved in the esteem of the world. It is almost god-like, power like that.

Well, can I be faced with the possibility of assuming god-like power and not assume it at once? Of course not.

As it happens, I have been spending nearly two years writing a history of science, and in the course of writing it I could not help but grow more or less intimate with about a thousand scientists of all shapes and varieties.[1]

Why not, then, make a list of the "ten greatest scientists of history"? Why not, indeed?

I sat down, convinced that in ten seconds I could rattle off the ten best. However, as I placed the cerebral wheels in gear, I found myself quailing. The only scientist who, it seemed to me, indubitably belonged to the list and who would, without the shadow of a doubt, be on such a list prepared by anyone but a consummate idiot, was Isaac Newton.

But how to choose the other nine?

It occurred to me to do as one did with the Academy Awards (and such-like affairs) and set up nominations, and after some time at that I found I had no less than seventy-two scientists whom I could call "great" with an absolutely clear conscience. From this list I could then slowly and by a process of gradual elimination pick out my ten best.

This raised a side issue. I would be false to current American culture if I did not give the ten winners a named award. The motion picture has its Oscar, television its Emmy, mystery fiction its Edgar, and science fiction its Hugo. All are first names and the latter two honor great men in the respective fields: Edgar Allan Poe and Hugo Gernsback.

For the all-time science greats, then, why not an award named for the greatest scientist of them all—Newton. To go along with the Oscar, Emmy, Edgar, and Hugo, let us have the Isaac. I will hand out Isaac Awards and choose the Isaac winners.[2]

Here, then, is my list of nominees, with a few words intended to indicate, for each, the reasons for the nomination. These are presented in alphabetical order—and I warn you the choice of nominees is entirely my own and is based on no other authority.

1. *Archimedes* (287?–212 B.C.). Greek mathematician. Considered the greatest mathematician and engineer of ancient times. Discovered the principle of the lever and the principle of buoyancy. Worked out a

---

[1] *The history referred to here has since been published as* Asimov's Biographical Encyclopedia of Science and Technology (*Doubleday, 1964*). *A revised edition appeared in 1972.*

[2] *If anyone has some wild theory that the choice of the name derives from any source other than Newton, let him try to prove it. Besides, what conceivable alternate origin could there be?*

ISAAC NEWTON

Newton was born in Woolsthorpe, Lincolnshire, on December 25, 1642, but he was a Christmas baby only by the Julian calendar then in use in England. By our modern calendar, he was born on January 4, 1643.

Newton, considered by many, if not most, historians of science, to be the greatest scientist who ever lived, almost didn't. He was born posthumously and prematurely and at first was not expected to survive.

He did, though, and proved a strange boy, interested in constructing mechanical devices of his own design, but showing no signs of unusual brightness. He seemed rather slow in his studies until well into his teens and apparently began to stretch himself only to beat the class bully, who happened to be first in studies as well.

good value for π by the principle of exhaustion, nearly inventing calculus in the process.

2. *Aristotle* (384–322 B.C.). Greek philosopher. Codified all of ancient knowledge. Classified living species and groped vaguely toward evolutionary ideas. His logic proved the Earth was round and established a world system that was wrong, but that might have proved most fruitful if succeeding generations had not too slavishly admired him.

3. *Arrhenius, Svante A.* (1859–1927). Swedish physicist and chemist. Established theory of electrolytic dissociation, which is the basis of modern electrochemistry. Nobel Prize, 1903.

4. *Berzelius, Jöns J.* (1779–1848). Swedish chemist. Was the first to establish accurate table of atomic weights. Worked out chemical symbols still used in writing formulas. Pioneered electrochemistry and notably improved methods of inorganic analysis.

5. *Bohr, Niels* (1885–1962). Danish physicist. First to apply quantum theory to atomic structure, and demonstrated the connection between electronic energy levels and spectral lines. Suggested the dis-

He graduated from Cambridge University in 1665 without particular distinction. Then the plague hit London and he retired to his mother's farm to remain out of danger. There his ideas simply exploded. He had already worked out the binomial theorem and was getting glimmerings of what came to be the calculus. What's more, seeing an apple fall from a tree in the distance (it did not hit him on the head) and noting the Moon in the sky, he wondered idly why the Moon did not fall and that led him to work out the mathematics of the manner in which the Moon did fall while also moving at right angles to that fall so that it remained in orbit about the Earth, though falling all the time.

It took him years to work out the mathematics of universal gravitation thoroughly, for he needed a correct size for the Earth and he had to complete the working out of calculus. Finally, in 1687, he published his Principia Mathematica, *the greatest single scientific work in history. There he presented his laws of motion and of gravitation in full detail.*

In 1692 he had a nervous breakdown and the rest of his life was rather subdued, but he had done enough (he had established the science of optics, built the first reflecting telescope, etc.). He died in London on March 20, 1727.

tribution of electrons among "shells" and rationalized the periodic table of elements. Nobel Prize, 1922.

6. *Boyle, Robert* (1627–1691). Irish-born British physicist and chemist. First to study the properties of gases quantitatively. First to advance operational definition of an element.

7. *Broglie, Louis V. de* (1892–   ). French physicist. Discovered the wave nature of electrons, and of particles in general, completing the wave-particle duality. Nobel Prize, 1929.

8. *Cannizzaro, Stanislao* (1826–1910). Italian chemist. Established usefulness of atomic weights in chemical calculations, and in working out the formulas of organic compounds.

9. *Cavendish, Henry* (1731–1810). English physicist and chemist. Discovered hydrogen and determined the mass of the Earth. Virtually discovered argon and pioneered in the study of electricity.

10. *Copernicus, Nicolaus* (1473–1543). Polish astronomer. Enunciated heliocentric theory of the solar system, with sun at center and Earth moving about it as one of the planets. Initiated the Scientific Revolution in the physical sciences.

11. *Crick, Francis H. C.* (1916–   ). English physicist and biochemist. Helped work out the helical structure of the DNA molecule,

ROBERT BOYLE

Boyle, born at Lismore Castle, Ireland, on January 25, 1627, the fourteenth child of the Earl of Cork, was an infant prodigy. He went to Eton at eight, at which time he was already speaking Greek and Latin, and traveled through Europe (with a tutor) when he was eleven. While in Geneva he was frightened by an intense thunderstorm into a devoutness that persisted for the rest of his life.

He was one of those who attended the periodic meetings of gentlemen scholars out of which the Royal Society arose. He heard of experiments on the production of vacuums and developed an air pump that could produce the best vacuum yet achieved mechanically.

He was one of the first to make use of an evacuated cylinder for a sealed mercury thermometer. He also produced an evacuated cylinder in which he showed that a feather and a lump of lead in the absence of air resistance would fall at the same speed. He also showed that sound could not be produced in a vacuum.

All this led him to experiment with gases. He found that by placing pressure on a gas he could force it into a smaller volume and that the pressure and volume were inversely proportional. This is still called "Boyle's law," and it was an important step in the formulation of an atomic theory.

In 1661 Boyle published The Sceptical Chemist, *which abandoned the ancient view of elements as mystical basic substances that could be reasoned out. Instead, he suggested that elements be defined experimentally. Anything that could not be broken down into simpler substances was an element. That was an important steppingstone toward the foundation of modern chemistry.*

His interest in religion grew and he would not serve as president of the Royal Society because he disapproved the form of the oath. He died in London on December 30, 1691, and in his will founded the Boyle Lectures, not on science but on the defense of Christianity against unbelievers.

which was the key breakthrough in modern molecular biology. Nobel Prize, 1962.

12. *Curie, Marie S.* (1867–1934). Polish-French chemist. Her investigations of radioactivity glamorized the subject. Discovered radium. Nobel Prize, 1903 (Physics) *and* 1911 (Chemistry). First person in history to win two.

13. *Cuvier, Georges L. C. F. D.* (1769–1832). French biologist. Founder of comparative anatomy and, through systematic studies of fossils, founder of paleontology as well.

14. *Dalton, John* (1766–1844). English chemist. Discovered law of multiple proportions in chemistry, which led him to advance an atomic theory that served as the key unifying concept in modern chemistry.

15. *Darwin, Charles R.* (1809–1882). English naturalist. Worked out a theory of evolution by natural selection which is the central, unifying theme of modern biology.

16. *Davy, Humphry* (1778–1829). English chemist. Established importance of electrochemistry by utilizing an electric current to prepare

CHARLES ROBERT DARWIN

Charles Robert Darwin was born in Shewsbury, Shropshire, on February 12, 1809, on the same day that Abraham Lincoln was born four thousand miles away in Kentucky. Darwin was the son of a well-to-do physician, the grandson of an even more famous one, Erasmus Darwin, and also grandson of Josiah Wedgwood, famous for his porcelainware.

Darwin showed no particular promise in his youth and was revolted by medicine (in the days before anesthesia) and would have nothing to do with it. He grew interested in natural history, however, and in 1831 took the post of ship's naturalist on the H.M.S. Beagle, which was setting off on a voyage of scientific exploration.

What followed was a five-year cruise around the world in which Darwin suffered agonies of seasickness and permanently impaired his health by contracting trypanosomiasis. In the course of the trip, Darwin carefully studied the species he encountered on the coasts of South America and on the islands of the Pacific. On the Galápagos Islands he studied fourteen species of finches unique to those islands. While wondering about so many species in so small an area, he began to get an inkling of the effect of isolation and natural selection.

In 1838 he read a book by T. R. Malthus on population and began to appreciate the role of population pressure on evolution. Carefully, he began to work out all the nuances of evolution by natural selection and spent years in writing and rewriting. Eventually (as friends had predicted) another biologist, A. R. Wallace, nearly anticipated Darwin, and the latter was forced to hasten into print. In 1859 his Origin of Species appeared and proved to be the foundation of modern biology.

Despite the controversy that followed and the obloquy hurled at him, Darwin, when he died in Down, Kent, on April 19, 1882, was honored as one of England's great men by being buried in Westminster Abbey.

elements not previously prepared by ordinary chemical means. These included such elements as sodium, potassium, calcium, and barium.

17. *Ehrlich, Paul* (1854–1915). German bacteriologist. Pioneered in the staining of bacteria. Worked out methods of disease therapy through immune serums and also discovered chemical compounds specific against diseases, notably syphilis. Hence founder of both serum therapy and chemotherapy. Nobel Prize, 1908.

18. *Einstein, Albert* (1879–1955). German-Swiss-American physicist. Established quantum theory, earlier put forth by Planck, by using it to explain the photoelectric effect. Worked out the theory of relativity to serve as a broader and more useful world-picture than that of Newton. Nobel Prize, 1921.

19. *Faraday, Michael* (1791–1867). English chemist and physicist. Advanced the concept of "lines of force." Devised the first electric generator capable of converting mechanical energy into electrical energy. Worked out the laws of electrochemistry and pioneered in the field of low-temperature work.

20. *Fermi, Enrico* (1901–1954). Italian-American physicist. Investigated neutron bombardment of uranium, initiating work that led to the atomic bomb, in the development of which he was a key figure. Outstanding theoretician in the field of subatomic physics. Nobel Prize, 1938.

21. *Franklin, Benjamin* (1706–1790). American universal talent. Demonstrated the electrical nature of lightning and invented the lightning rod. Enunciated the view of electricity as a single fluid, with positive charge representing an excess and negative charge a deficiency.

22. *Freud, Sigmund* (1856–1939). Austrian neurologist. Founder of psychoanalysis and revolutionized concepts of mental disease.

23. *Galileo* (1564–1642). Italian astronomer and physicist. Studied the motion of falling bodies, disrupting the Aristotelian world system and laying the foundation for the Newtonian one. He popularized experimentation and quantitative measurement and is the most important single founder of experimental science. He was the first to turn a telescope upon the heavens and founded modern astronomy.

24. *Gauss, Karl F.* (1777–1855). German mathematician and astronomer. Perhaps greatest mathematician of all time. In science, developed method of working out planetary orbit from three observations and made important studies of electricity and magnetism.

25. *Gay-Lussac, Joseph L.* (1778–1850). French chemist and physi-

cist. Discovered several fundamental laws of gases and was the first to ascend in balloon to make scientific measurements at great heights.

26. *Gibbs, Josiah W.* (1839–1903). American physicist and chemist. Applied principles of thermodynamics to chemistry and founded, in detail, chemical thermodynamics, which is the core of modern physical chemistry.

27. *Halley, Edmund* (1656–1742). English astronomer. First to undertake systematic study of southern stars. Worked out the orbits of comets and showed that they were subject to the law of gravitation.

28. *Harvey, William* (1578–1657). English physiologist. First to apply mathematical and experimental methods to biology. Demonstrated the circulation of the blood, overthrowing ancient theories and founding modern physiology.

29. *Heisenberg, Werner* (1901–   ). German physicist. Enunciated uncertainty principle, a concept of great power in modern physics. Was the first to work out the proton-neutron structure of the atomic nucleus and was thus the founder of modern nucleonics. Nobel Prize, 1932.

30. *Helmholtz, Hermann L. F. von* (1821–1894). German physicist and physiologist. Advanced a theory of color vision and one of hearing, making important studies of light and sound. First to enunicate, clearly and specifically, the law of conservation of energy.

31. *Henry, Joseph* (1797–1878). American physicist. Devised first large-scale electromagnet and invented electric relay, which was basis of the telegraph. Invented the electric motor, which is the basis of much of modern electrical gadgetry.

32. *Herschel, William* (1738–1822). German-English astronomer. Discovered the planet Uranus, first to be discovered in historic times. Founded the modern study of stellar astronomy by work on double stars, on proper motions, etc. He was the first to attempt to work out the general shape and size of the Galaxy.

33. *Hertz, Heinrich R.* (1857–1894). German physicist. Discovered radio waves, thus establishing Maxwell's theoretical predictions concerning the electromagnetic spectrum.

34. *Hipparchus* (second century B.C.). Greek astronomer. The greatest of the naked-eye observers of the heavens. Worked out the epicycle theory of the solar system, with the Earth at the center. Perfected system of latitude and longitude, devised first star map, and discovered the precession of the equinoxes.

ALBERT EINSTEIN

Einstein, although Jewish, was educated in a Catholic grammar school in Munich, Bavaria. His family had moved there from nearby Ulm, where Einstein had been born on March 14, 1879. He was so slow in learning at first that there was some feeling he might prove retarded. In high school he was interested only in mathematics and was invited to leave by a teacher who said, "You will never amount to anything."

He managed to get into college in Switzerland, but with difficulty, for he qualified only in mathematics. He cut the lectures while reading up on theoretical physics and passed only by using the excellent lecture notes compiled by a friend. After graduation, he could do nothing better than get a job as a junior official at the Patent Office at Berne.

There, without a laboratory and without academic connections, he revolutionized the world of science. The year 1905 saw the publication of papers in which he accomplished three major feats. One paper dealt with the photoelectric effect, whereby light falling upon certain metals was found to stimulate the emission of electrons. Einstein explained it by quantum theory which had been advanced five years earlier by Max Planck. It was Einstein's use of it that really established it.

In another paper, he worked out the mathematical analysis of Brownian motion, which led to the first real notion of the actual size of atoms and molecules. And then, too, of course, he worked out the special theory of relativity in the same year—following it ten years later with the general theory of relativity. He received the 1921 Nobel Prize in physics for his work on the photoelectric effect.

When Hitler came to power in Germany, Einstein, who had lived there since 1913, was forced to leave and take up permanent residence in the United States. Here he was persuaded by Leo Szilard and others to write a letter about German successes in atomic research to President Roosevelt in 1939. This led to the setting up of the Manhattan Project and began American work toward the atomic bomb. He died in Princeton, New Jersey, on April 18, 1955.

35. *Hubble, Edwin P.* (1889–1953). American astronomer. His studies of the outer galaxies demonstrated that the universe was expanding. Presented first picture of known universe as a whole.

36. *Hutton, James* (1726–1797). Scottish geologist. Founded modern geology; the first to stress the slow, eons-long, changes of the Earth's crust under environmental stresses continuing and measurable in the present.

37. *Huygens, Christian* (1629–1695). Dutch mathematician, physicist, and astronomer. Devised first pendulum clock, thus founding the art of accurate timekeeping. Improved the telescope and discovered Saturn's rings. Was the first to advance a wave theory of light.

38. *Kekule von Stradonitz, Friedrich A.* (1829–1896). German chemist. Devised the modern method of picturing organic molecules with bonds representing valence links, of which the carbon atom possessed four. This brought order into the jungle of organic chemistry.

39. *Kelvin, William Thomson, Lord* (1824–1907). Scottish physicist. Proposed absolute scale of temperature, did important theoretical work on electricity, and was one of those who worked out the concept of entropy.

40. *Kepler, Johannes* (1571–1630). German astronomer. Established elliptical nature of planetary orbits, and worked out generalizations governing their motions. He thus established the modern model of the solar system and eliminated the epicycles that had governed astronomical thinking for nearly two thousand years.

41. *Kirchhoff, Gustav R.* (1824–1887). German physicist. Applied the spectroscope to chemical analysis, thus founding modern spectroscopy and laying the groundwork for modern astrophysics. He was the first to study black-body radiation, something which led, eventually, to the quantum theory.

42. *Koch, Robert* (1843–1910). German bacteriologist. Isolated bacteria of tuberculosis and of anthrax. Was the first to develop systematic methods for culturing pure strains of bacteria and established rules for locating the infectious agent of a disease. Nobel Prize, 1905.

43. *Laplace, Pierre S. de* (1749–1827). French mathematician and astronomer. Worked out the gravitational mechanics of the solar system in detail and showed it to be stable.

44. *Lavoisier, Antoine L.* (1743–1794). French chemist. First to popularize quantitative methods in chemistry. Established the nature of combustion and the composition of the atmosphere. Enunciated

the law of conservation of matter. Introduced the modern system of terminology for naming chemical compounds and wrote the first modern chemical textbook.

45. *Lawrence, Ernest O.* (1901–1958) American physicist. Invented the cyclotron, first device suitable for induction of large-scale artificial nuclear reactions. Modern nuclear-physics technology depends upon the cyclotron and its descendants. Nobel Prize, 1939.

46. *Leverrier, Urbain J. J.* (1811–1877). French astronomer. Worked out the calculations that predicted the position of the then-unknown Neptune. Its subsequent finding was the greatest victory for gravitational theory and the most dramatic event in the history of astronomy.

47. *Liebig, Justus von* (1803–1873). German chemist. Worked out methods of quantitative analysis of organic compounds. Was the first to study chemical fertilizers intensively and hence is the founder of agricultural chemistry.

48. *Linnaeus, Carolus* (1707–1778). Swedish botanist. Painstakingly classified all species known to himself into genera, placed related genera into orders and related orders into classes, thus founding taxonomy. He devised the system of binomial nomenclature, whereby each species has a general and a specific name.

49. *Maxwell, James Clerk* (1831–1879). Scottish physicist. Worked out equations that served as basis for an understanding of electromagnetism. Showed light to be an electromagnetic radiation and predicted a range of such radiations beyond those then known. Worked out the kinetic theory of gases, one of the foundation blocks of physical chemistry.

50. *Mendel, Gregor J.* (1822–1884). Austrian botanist. His studies of pea plants founded the science of genetics, though the laws of inheritance he worked out remained unknown in his lifetime.

51. *Mendeleev, Dmitri I.* (1834–1907). Russian chemist. Worked out the periodic table of the elements, which proved an important unifying concept in chemistry. The value of the table was established by his dramatic prediction of the properties of as-yet-unknown elements.

52. *Michelson, Albert A.* (1852–1931). German-American physicist. Made accurate determinations of velocity of light. Invented the interferometer and used it to show that light travels at constant velocity in all directions despite motion of the earth. This served as the foundation of the theory of relativity. Nobel Prize, 1907.

53. *Moseley, Henry G. J.* (1887–1915). English physicist. Studied X-ray emission by elements and worked out the manner in which nuclear electric charge differed from element to element. This led to the concept of the atomic number, which greatly improved the rationale behind the periodic table of the elements.

54. *Newton, Isaac* (1642–1727). English physicist and mathematician. Invented calculus, thus founding modern mathematics. Discovered compound nature of white light, thus founding modern optics. Constructed the first reflecting telescope. Worked out the laws of motions and the theory of universal gravitation, replacing Aristotle's world system with one that was infinitely better.

55. *Ostwald, Friedrich Wilhelm* (1853–1932). German physical chemist. Founder of modern physical chemistry. Worked on electrolytic dissociation. Proposed the modern view of catalysis as a surface phenomenon. Nobel Prize, 1909.

56. *Pasteur, Louis* (1822–1895). French chemist. Did pioneer work in stereochemistry. Advanced the germ theory of disease, thus founding modern medicine. He worked out dramatic methods of inoculation against various diseases.

57. *Pauling, Linus C.* (1901–   ). American chemist. Applied quantum theory to molecular structure, proposing a new and more useful view of the valence bond, and establishing modern theoretical organic chemistry. First to propose the helical structure of large organic molecules, such as proteins, which led on to Crick's work. Nobel Prize, 1954 (Chemistry) *and* 1963 (Peace). Second person to win two Nobel Prizes.

58. *Perkin, William H.* (1838–1907). English chemist. Initiated the great days of synthetic organic chemistry by synthesizing aniline purple, first of the aniline dyes. Also synthesized coumarin, founding the synthetic perfume industry.

59. *Planck, Max K. E. L.* (1858–1947). German physicist. Worked out quantum theory to explain the nature of black-body radiation. This theory treats energy as discontinuous and as consisting of discrete particles or quanta. The new understanding it offered is so crucial that physics is commonly divided into "classical" (before Planck) and "modern" (since Planck). Nobel Prize, 1918.

60. *Priestley, Joseph* (1733–1804). English chemist. Discovered oxygen.

61. *Roentgen, Wilhelm K.* (1845–1923). German physicist. Discov-

ered X rays, an event usually considered as initiating the Second Scientific Revolution. Nobel Prize, 1901.

62. *Rutherford, Ernest* (1871–1937). New Zealand-born British physicist. Enunciated the theory of the nuclear atom, in which the atom was viewed as containing a tiny central nucleus surrounded by clouds of electrons. This founded subatomic physics. Rutherford was the first to effect an artificial nuclear reaction, changing one element into another. Nobel Prize, 1908.

63. *Scheele, Carl W.* (1742–1786). German-Swedish chemist. Discovered or co-discovered some half-dozen elements, as well as a variety of organic and inorganic compounds.

64. *Schwann, Theodor* (1810–1882). German zoologist. Discovered first animal enzyme, pepsin. Contributed to the disproof of spontaneous generation. Strongest single contributor to the establishment of the cell theory, which is virtually the atomic theory of biology.

65. *Soddy, Frederick* (1877–1956). English chemist. Worked out the isotope theory of the elements and with it the details of the course of radioactive breakdown. Nobel Prize, 1921.

66. *Thales* (640?–546 B.C.). Greek philosopher. Founder of rationalism and the tradition that has led to modern science.

67. *Thomson, Joseph J.* (1856–1940). English physicist. First to establish, definitely, that cathode rays consisted of particles far smaller than atom; therefore the discoverer of the electrons and the founder of the study of subatomic particles. Nobel Prize, 1906.

68. *Van't Hoff, Jacobus H.* (1852–1911). Dutch physical chemist. Advanced theory of the tetrahedral carbon atom, by which molecular structure could be described in three dimensions. Contributed greatly to chemical thermodynamics. Nobel Prize, 1901.

69. *Vesalius, Andreas* (1514–1564). Belgian anatomist. Described his anatomical observations in a book with classically beautiful illustration. This demolished ancient errors in anatomy and established the science in its modern form. Published in 1543, it began the Scientific Revolution in the biological sciences.

70. *Virchow, Rudolf* (1821–1902). German pathologist. Studied disease from the cellular standpoint and ranks as the founder of modern pathology. He also labored on behalf of sanitation reform and was one of the founders of modern hygiene.

71. *Volta, Alessandro* (1745–1827). Italian physicist. Built the first chemical battery and founded the study of current electricity.

72. *Wöhler, Friedrich* (1800–1882). German chemist. First to form an organic compound (urea) from an inorganic precursor, thus founding modern organic chemistry.[3]

Having completed the list of nominees, I am under the temptation to play with it, analyze it statistically in various fashions. I shall succumb to this in only one small way. Let me list in Table 35 the total number of scientists on the list according (as nearly as I can guess) to the language they thought in.

Table 35

English	26
German	21
French	7
Italian	4
Greek	4
Swedish	3
Dutch & Flemish	3
Polish	2
Danish	1
Russian	1

I suppose this can be taken as evidence that modern science is primarily an Anglo-American-German phenomenon. I think, though, it is more likely to demonstrate that the individual who selected the names is himself English-speaking.[4]

Now there is nothing left for me to do but to list my version of the ten winners of the Isaac Awards. It will be placed on another page[5] so that you can prepare your own version (if you choose) without reference to mine. My own list of Isaac winners is in alphabetical order; for I lack the courage to choose among them (except that I would put Newton first). However, you may list yours in a particular order, if you have the courage.

---

[3] *I repeat that this list is perhaps overconservative. Arguments can be advanced for including such men as Hippocrates, Euclid, Leonardo da Vinci, Robert H. Goddard, Charles H. Townes, Emil Fischer, and so on.*

[4] *Or maybe not, considering the similar results obtained from a consideration of the Nobel Prizes, see the previous chapter.*

[5] *No, it won't. I did it in the original book and then it was left out in the paperback edition and I had to engage in a lot of correspondence. You will find it here, then, in the table immediately following.*

Please feel free to send your list to me, if you wish. You have the right to disagree and can freely tell me about the men I have included (or excluded) that only a jackass would include (or exclude), either in my list of nominees or in my final list of Isaac winners.

It's possible you may even enlighten me and cause me to change my mind.

Table 36—The Isaac Winners

Archimedes
Darwin, Charles R.
Einstein, Albert
Faraday, Michael
Galileo
Lavoisier, Antoine L.
Maxwell, James Clerk
Newton, Isaac
Pasteur, Louis
Rutherford, Ernest

Of these ten winners, five are British, two are French, and there is one Greek, one Italian, and one German. There are five Protestants, three Catholics, one Jew, and one pagan.

Since this article was written in 1963, I have not yet been tempted to alter the list despite many letters I have received on the subject.

INDEX